An Engineers Workshop 1860

A HISTORY OF
MACHINE TOOLS

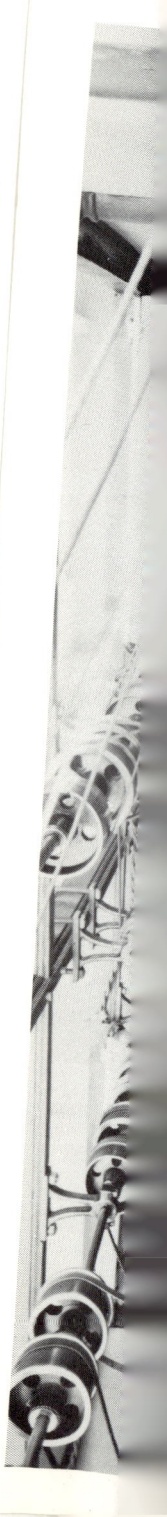

A HISTORY OF
MACHINE TOOLS

IAN BRADLEY

Sometime in charge of the
Research and Development Workshops
Vickers-Armstrong Limited,
South Marston, Swindon, Wilts.

Model and Allied Publications Limited
13/35 Bridge Street, Hemel Hempstead, Hertfordshire

Model and Allied Publications Limited
13/35 Bridge Street
Hemel Hempstead
Hertfordshire

First Published 1972

© Model and Allied Publications Ltd. 1972

ISBN 0 852 42102 8

Printed in England by Page Bros (Norwich) Ltd.
Bound in England by Mansells (Bookbinders) Ltd.

iv

CONTENTS

v

ACKNOWLEDGEMENTS

I am greatly indebted to all those who encouraged the researches that culminated in the writing of this book. Their help on many points of detail has been of great assistance.

Many firms have provided me with illustrations and information that have proved invaluable; I am particularly in debt to the following who have given unsparingly of their resources:

The Amalgamated Union of Engineering and Foundry Workers. James Archdale. The Butler Machine Tool Co. Limited. Dean Smith & Grace Limited. T. S. Harrison & Sons Limited. The Jacobs Manufacturing Co. Limited. Kearney & Trecker Limited. Myford Limited. W. J. Meddings Limited. James Neill & Co. (Sheffield) Ltd. Perfecto Engineering. Staveley Machine Tools Limited. Vickers Limited. Wadkin Limited.

ILLUSTRATION ACKNOWLEDGEMENTS

Crown Copyright, Science Museum, London.
Frontispiece, p. xiv C and D, Figures 1.3, 1.8, 1.11, 8.4, 10.6, 10.16, 10.17 and 12.4.

Photo Science Museum, London
Figure 14.2

Lent to the Science Museum by James Barr.
Figure 10.24

PREFACE

AT ONE TIME the development of machine tools was of historical interest only to a few specialists. To-day, however, when the fruits of these tools are plain for all to see, there must be many people, and not necessarily all with a purely engineering training, who would welcome information concerning the background from which modern machine tool technology has sprung.

At a time when the interest in antiques has reached such widespread proportions it is surprising that more is not heard or seen, except perhaps in isolated museums, of the beautiful machines produced by clockmakers and others in the 18th century. These must have given inspiration to many of the world's toolmakers and may have been the basis of much of the work undertaken in the early stages of development.

Whilst it has been possible to cover the period from 1700 to 1900 in some detail it must be appreciated that technology tends to gather momentum, especially under the influence of military activity, so the advances made from the turn of the century until the present day must necessarily be an abridged resumé of progress in this field. To do otherwise would involve the production of a further book of some complexity, of value probably to only a narrow spectrum of readers. It is hoped, however, that the present work may provide an outline that will stimulate an increased interest in the subject.

I.B.   *Wickham, Berkshire, 1972*

FOREWORD

IN ENGLAND intensive development of machine tools took place all within a space of some 150 years, the men associated with it being either employees or partners in firms engaged in engineering pursuits.

The period covered is from about 1725 until 1875 and the tools involved, for the most part, were those that filled the needs of the developer himself in his own particular field of engineering. For it was not until the Great Exhibition of 1851, when Joseph Whitworth staged a display of their unsurpassed machine tools, that it was realised how important the role of the specialist manufacturer could be and how lucrative his future might prove.

Over the years those interested formed a closely knit federation, so much that it is possible to devise a genealogical tree showing the connection of the most important names one with another.

Thus Bramah, who started as a cabinet maker, employed Maudslay who in turn became associated with Whitworth, Roberts, Brunell and Nasmyth, either in an advisory capacity of directly developing machines to further their interests.

It was, however, left to Whitworth to realise how important, in the future, technology was likely to prove, and to take steps that would help to ensure a supply of the best brains being available. This he did by founding the Whitworth Scholarships setting aside an annual sum of £300 for the purpose.

In order to save reiteration during the course of the book it has been thought desirable, at the outset, to include a potted biography of each of the principal actors in the development period to which reference has been made.

These notes can conveniently be read in with the diagram showing the association of the major participants in the period under review.

By way of explanation it should be noted that the letter 'E' opposite an arrow head on the line connecting two names indicates that the man

Foreword

whose name appears *above* the arrow employed the person to whom
the arrow points either directly or in association.

It has also been thought convenient, for those who wish to know
more about the men involved, to give details of references in *Encyclopaedia Brittanica, 9th, 10th,* and *11th Editions* or elsewhere, where a
full biography can be found in each case.

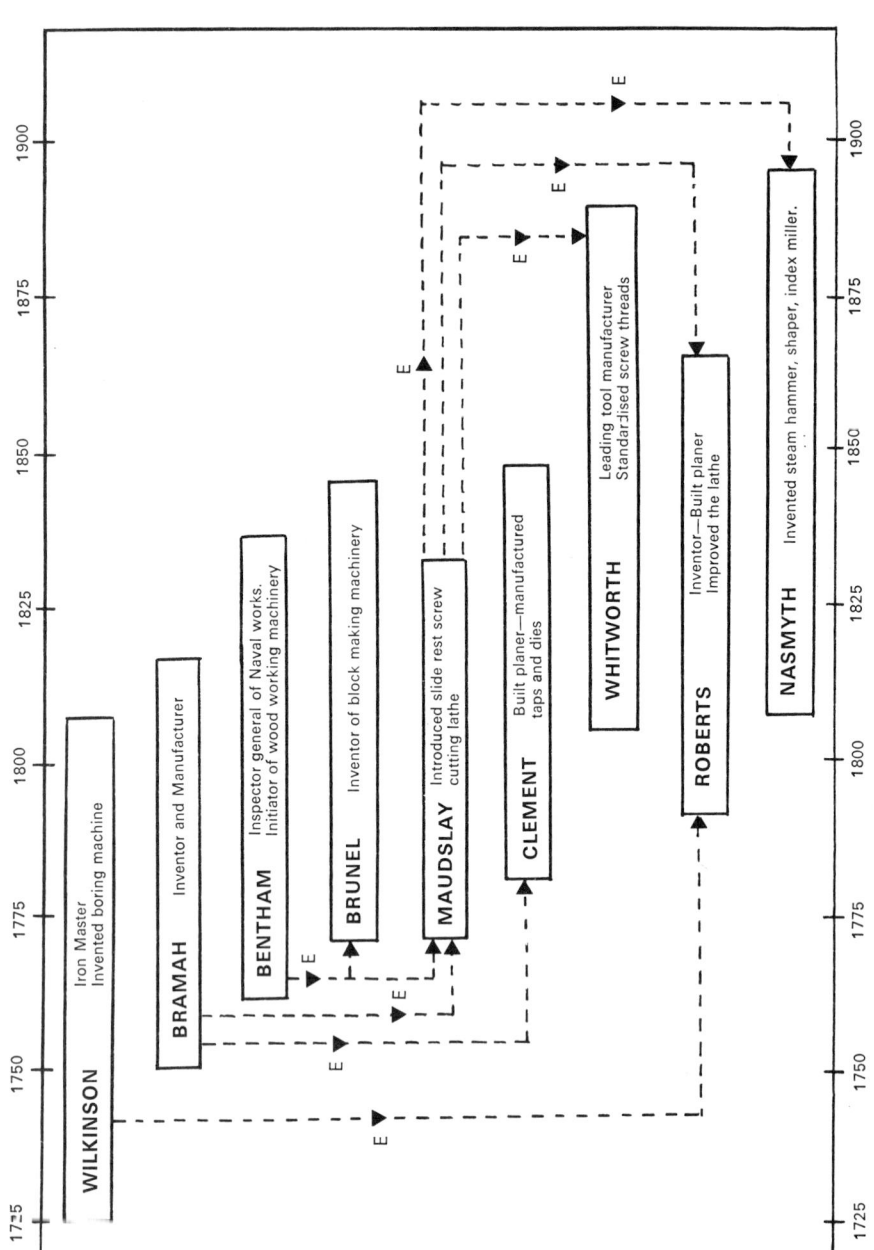

x

BIOGRAPHIES

BENTHAM
Encyclopaedia Brittanica, Vol. 7, 314B.

'General' Bentham, was responsible for the construction and supervision of H.M. Dockyards. Perhaps the most important of these was, at the time of publication 1877, situated at Chatham "where every item of turnery, etc., is performed for the Navy, from boring the chamber of a pump to the turning of a button for a chest of drawers".

It was at Chatham that Brunel's remarkable block making machinery was set up under Bentham's direction.

A second set was also installed in Portsmouth Dockyard.

The plant at Chatham, however, was never used since Portsmouth could supply all the blocks that the Navy required.

BRAMAH, Joseph, 1748–1814
Ency. Britt. Vol. 4, 213B, Vol. 18, 483C

Joseph Bramah was born at Stamborough in Yorkshire on April 13th 1749. When he was about sixteen he was incapacitated by an injury to his ankle and so was unable to follow his intended employment as an agricultural labourer.

Bramah, therefore, became apprenticed to a carpenter and joiner, later obtaining employment with a cabinet maker and eventually becoming a principal in that business.

Although initially having had a grounding in woodwork, his fertile brain led him into the field of pumps, fire engines, steam boilers and papermaking machinery for all of which he obtained patents. His most notable contribution to the machine tool industry was his invention of the hydraulic press, the later development of which was to have far-reaching results in any part of industry where immense forces were needed.

He was also responsible for the special type of lock that bears his name, and in this he was helped by one of his men, Henry Maudslay, who was instrumental in solving some of the complex machining problems involved in its manufacture.

xi

H. MAUDSLAY.

Henry Maudslay

Biographies

BRUNEL, M. I., 1769–1849
Ency. Britt. 9th & 10th Vol. 4, 396D, Vol. 3, 833C, Vol. 32, 540B

Brunel was born at Hagneville in Normandy in 1769. Having served for a period in the French Navy on the return from one of his voyages in 1792 he found the Revolution at its height, so, like many others with Royalist pursuasions, he was forced to seek safety in flight, emigrating to America.

The idea of making wood blocks by machiner had been maturing in his mind, so, in 1799, he visited England where he thought his proposals would be the more readily received. Having had a meeting with Earl St. Vincent, at that time head of the Admiralty, after many administrative delays and on the recommendation of Earl Spencer and General Bentham, the system for making these ship's blocks was adopted.

With great discrimination Henry Maudslay was selected by Brunel to make the equipment.

At a time when the *Great Britain*, the first screw-propelled iron-built ship, is in the news, having been returned to the place of her birth, it is interesting to recall that it was Brunel who had initiated and supervised her construction.

CLEMENT, Joseph, 1779–1849
Ency. Britt. 9th and 10th Vol. 15, 155B

Clement was a gifted mechanic who, in company with Maudslay, made many improvements to the lathe, notably in the provision of self-acting mechanism to lathes designed for facing operations. This he did in 1827.

He also produced, in 1825, the first planing machine resembling those made towards the end of the 19th century.

Clement was another member of Bramah's engineering coterie. He had started active life on the land, but, finding that he was more interested in mechanical matters, he pursuaded the local blacksmith to allow him access to his forge where he quickly became proficient in the use of hand tools.

His ability to use tools got him a post with a firm in the neighbourhood making power looms. But his experience here, and the greater indication he received on moving to Carlisle, convinced him that the men in the real money were the draughtsmen and designers. Accordingly he determined to train in draughtsmanship and having done so gravitated down South to London where his design qualifications secured him immediate employment, leading eventually to his becoming chief draughtsman for Maudslay Son and Field. Clement stayed

Richard Roberts' Planing Machine 1817

Wilkinson's Boring Mill

with them for many years assisting in the production of marine engines for which the firm had become famous. Eventually specialising in the making of machines for particular duties and carrying on with the improvements to screw threading equipment started by Maudslay.

Clement's shops were notable for the quality of their workmanship and, as subsequent events were to prove, for the calibre of the men employed there. As an example Joseph Whitworth was one of them.

MAUDSLAY, Henry, 1771–1831
Ency. Britt. Vol. 4, 398B, Vol. 15, 152D
'Engineers, Inventors and Workers', page 77 et seq.

Maudslay, was brought up near Woolwich Arsenal. His father, a cabinet maker, having joined the army and then been invalided out, became a storeman in the Arsenal. This enabled the young Maudslay to obtain a position in that organisation in which he rapidly became a proficient worker, particularly after he was transferred to the metal-working shops. His ability came to the ears of Bramah who was looking for a capable mechanic to assist him in the development of his patented lock. However, when Maudslay appeared at Bramah's works both Bramah and his foreman were at first doubtful of his capability for he had served no recognised apprenticeship. But on Maudslay offering to re-build a damaged bench vice as a test of his proficiency and finding that he did so expeditiously and well, they became convinced of his suitability for the post offered.

Maudslay stayed with Bramah for ten years eventually becoming foreman. He left to found his own business, having in the meantime married Sarah Tindel, Bramah's housemaid. His claim to fame in the field of machine tools lies in his invention of the screwcutting lathe and the slide rest. But his employment by Brunel to produce the woodworking machinery for the making of ships blocks established him as an engineering designer and manufacturer. The block-making plant, probably the first example of mass production, was a complete success and led to a great increase in business involving an expansion of workshops facilities, leading eventually to the formation of the firm Maudslay Sons and Field which was carried on by his eldest son after Maudslay's death in 1831

NASMYTH, James, 1808–1890
9th & 10th Ency. Britt. Vol. 31, 76B, Vol. 17, 238C, Vol. 9, 413B,
Vol. 11, 426A and Vol. 13, 328B

The steam hammer was patented by Nasmyth in 1842. Originally designed so that the hammer head was lifted when steam was applied and was then allowed to fall by its own weight, the steam hammer has been modified to admit steam above the piston and so increase the force of the hammer blow.

Every increase in the weight of a vertical steam hammer requires a very much larger increase in the weight and solidity of the anvil which resists the blow. A 40-ton hammer once in use in Woolwich Arsenal had an anvil weighing some 160 tons and there were nearly 500 tons of iron in its foundation.

See also Ency. Britt. Vol. 13, 328B and Samuel Smiles Autobiography

Nasmyth was born in Edinburgh on August 19th 1808. He went to school in his native centre, attending classes in chemistry, mathematics and natural philosophy at the University. He developed a considerable skill in the production of working models of such things as steam engines. These found a ready sale and were instrumental in securing Nasmyth a place in Henry Maudslay's workshop, where his ability to use tools was instantly appreciated.

Nasmyth spent two years with Maudslay, leaving his employment in 1830. In 1834 he started in business on his own account. The beginnings were small, but business quickly developed, to such an extent that by 1856, he was able to retire as a very rich man.

ROBERTS, Richard, 1789
Samuel Smiles 'Industrial Biography' page 264 etc. See also Encyc. Britt. Vol. 6, 448B

Roberts was born in 1789 at Carreghora in the parish of Llanymynech, Montgomeryshire. His father was a shoemaker who could give his son little education so, as soon as the boy was old enough he was put to common labouring work. For a time he worked in a quarry near his home, but finding little interest in the work he used to spend much of his time in making various pieces of mechanism, an occupation for which he was developing an aptitude.

He seems to have rapidly gained a facility in the use of hand tools, this encouraged him to seek employment in a neighbouring iron works under John Wilkinson, the famous ironmaster who had invented the boring machine that successfully machined the cylinders of Watt's steam engine

On leaving Wilkinson's works he went to Birmingham passing through several engineering establishments in the years that followed. In this way he gained a wide experience in mechanical processes. Unfortunately, at the end of this period, Roberts was drawn by ballot for service in the militia so he determined to make his way to London where the authorities would find it hard to trace him. In London he joined Maudslay's workshop where his fellow workers, if not all of them contemporaries, eventually became very well known in engineering circles.

By 1816 Roberts was in business for himself. The list of his inventions is formidable, many of them applicable to the cotton industry for which he was probably the first of the machine makers.

At the time when Samuel Smiles was writing about him (1878) Richard Roberts may still have been alive, but he would have been an old man approaching his 90th year. Smiles, however, does not speak of his death but does tell us that his closing years were those of much impoverishment. Roberts was undoubtedly a great engineer, but he was a poor financier allowing others to exploit him and to steal ideas which, otherwise, would have made him a successful and wealthy man.

WILKINSON, The Iron Master, 1728–1808
9th and 10th Ency. Britt. Vol. 7, 77A

Wilkinson was the son of a day labourer in an iron furnace who rose to be overlooker or foreman. The father was eventually in business for himself, making a box-iron for the use of laundresses. This iron formed the basis of the family fortune after the father and son had joined forces.

But the most important of the younger Wilkinson's inventions was the boring machine, its success being proved by the boring of the cylinders of Watt's steam engine. The accuracy of the work so far exceeded anything previously available that it was largely responsible for the high reputation that Watt and his partner Boulton had in the Midlands.

So far as is known this boring machine is the only machine tool to be developed by Wilkinson himself. But he had surrounded himself with men of fertile and inventive mind, amongst whom was Roberts, the inventor of the planing machine.

WHITWORTH, Sir Joseph, Bart., 1803–1887

Whitworth was born at Stockport near Manchester on December the 21st, 1803.

He was at school until the age of 14 and on leaving was placed with his uncle, a cotton spinner, with a view to becoming a partner in the business.

His taste for mechanics, however, found no satisfaction in this work so he soon gave it up, spending some time with various engineering manufacturers near Manchester.

In 1825 he moved to London where he gained further experience in machine shops, notably those of Henry Maudslay.

In 1833 he returned to Manchester and set up in business as a toolmaker. From then on his business prospered until, on his death in 1887, his residuary legatees were able to provide no less than £100,000 for the permanent establishment of the 30 Whitworth Scholarships he had founded in 1868, as well as half a million to charitable objects.

CHAPTER ONE

THE LATHE

OF ALL THE TOOLS in the armoury of the production engineer undoubtedly the lathe in its many forms is the most important. It is also probably the oldest, since elementary forms of the lathe appear to have been in use from the 14th century onward.

The pole lathe, illustrated in Fig. 1.1, was used very early on to turn chair legs and other similar parts, and was often found actually in the woods from whence the raw material was drawn, because an essential ingredient in the construction or make-up of the pole lathe, namely a strong, springy sapling, was available on the spot. It will be appreciated that work, mounted between centres in the manner shown, and driven by means of a cord wound around it, does not revolve continuously but moves backwards and forwards; forwards when the operative presses down the treadle lever and backwards when he releases it to allow the sapling to reverse the motion and return the treadle to its original position ready for the next working stroke.

With indoor versions of the pole lathe, similarly to that in the illustration, the sapling was replaced by a bow fixed to the ceiling of the workshop.

Inevitably, the discontinuous rotation of the work inseparably from the arrangement became intolerable and some means of improving matters had to be sought. These resulted in the introduction of the wheel lathe, an example belonging to John Smeaton of lighthouse fame, being seen in the illustration Fig. 1.2. The Wheel Lathe needed two men to operate it, one to provide the motive power the other to do the actual turning operations. An enlarged version of the Wheel Lathe is depicted in Fig. 1.3. Here, the Great Wheel Lathe from the wheelwright's shop in Barley, Hertfordshire is seen in a reconstruction of the premises. The Lathe seen in the illustration was in use as late as 1949, having been worked by four generations of Wheelwrights. It was used to turn the elm hubs for wagon wheels and its proportions were quite impressive. The flywheel was over 6ft in diameter and to it was bolted a driving pulley, some 3ft across, which the assistant turned by means of a handle. The drive to the mandrel was by a rope. In early lathes, initially, the work was mounted between dead centres, that is a pair of centres neither of which rotate. A pulley was attached directly to the work as seen in 'A', Fig. 1.4, and the driving belt was often crossed to avoid slip. A further development in the driving arrangements removed the pulley attached directly to the work and transferred it to one of the dead centres upon which it ran. the work itself being driven by a pin

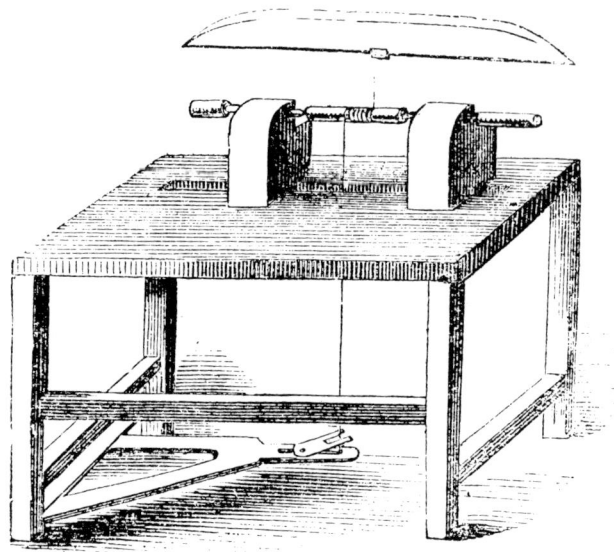

Fig. 1.1.   The Pole Lathe.

fixed to the pulley through the medium of a carrier as depicted at 'B' in the same illustration.

The further and final development provided the lathe with one live mandrel carrying the driving pulley, this last arrangement permitting work other than that set between centres to be mounted on the live mandrel itself. These mandrel arrangements are seen at 'C' in Fig. 1.4.

In common with similar lathes at the time Smeaton's lathe was constructed mainly of wood, but by the end of the 18th century iron and other metals began to be generally used in the manufacture of lathes. Consequently their capacity increased so that it was not uncommon for two men to be employed for turning the driving wheel.

Shortly after the First World War had ended in 1918 a device called the Verschoyle mandrel was put on the market. This little machine,

SMEATON'S LATHE.

Fig. 1.2   Smeaton's Lathe.

B

which is illustrated in Fig. 1.5 bore a strong likeness to the wheel lathe having inherited its greatest drawback in that two men were needed for its successful working. It was said that it could be used by a single operative; but turning the mandrel with one hand and endeavouring to guide a turning tool with the other must have been a pretty exasperating experience. At all events the device was not a success and soon went off the market.

Fig. 1.3   The Great Wheel Lathe.

But to return to the original wheel lathe. Its drawbacks soon led to the demand for a tool that could be worked by one man alone and this led to the introduction of the Treadle Lathe illustrated in Fig. 1.6. Here as will be seen, the treadle is connected, through a pitman, with a crankshaft supported usually on centres fitted to the stand carrying the lathe bed itself. The driving pulley consists of a heavy iron flywheel provided with grooves to accommodate a round catgut or leather belt used to transmit power to the mandrel. Whilst both the headstock and tailstock appear to be iron castings supporting the mandrels normally associated with them, it is not certain that the particular lathe illustrated has a cast-iron bed and legs. It some instances these components were still made of wood.

At this stage it is as well for the readers to acquaint themselves with the salient parts of the lathe and the names given to them. The

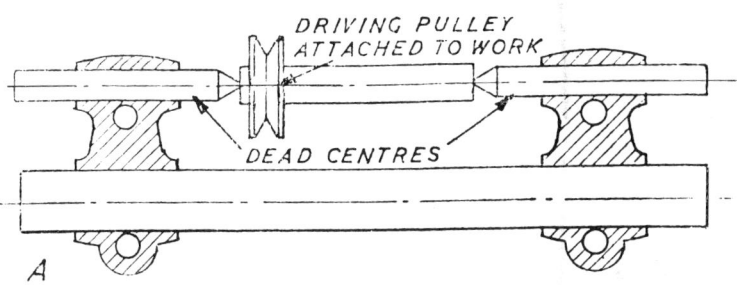

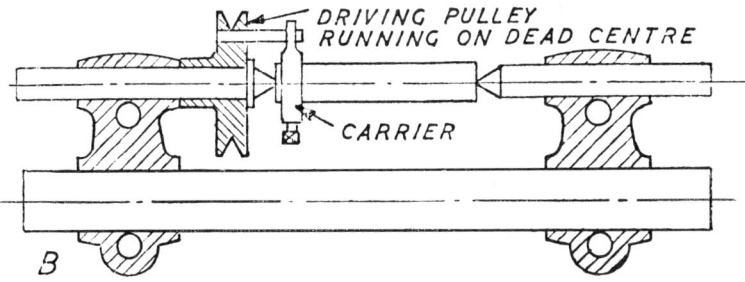

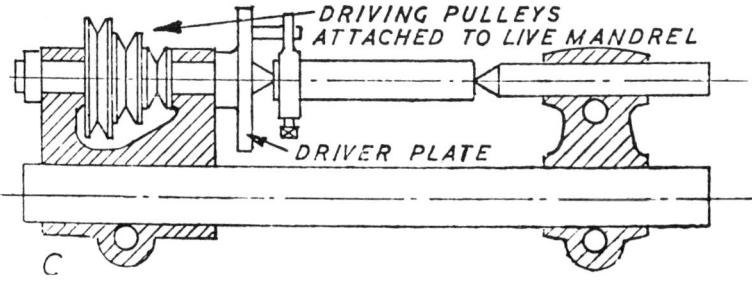

Fig. 1.4   Methods of driving the work.

Fig. 1.5   The Verschoyle Mandrel.

Fig. 1.6   The Treadle Lathe.

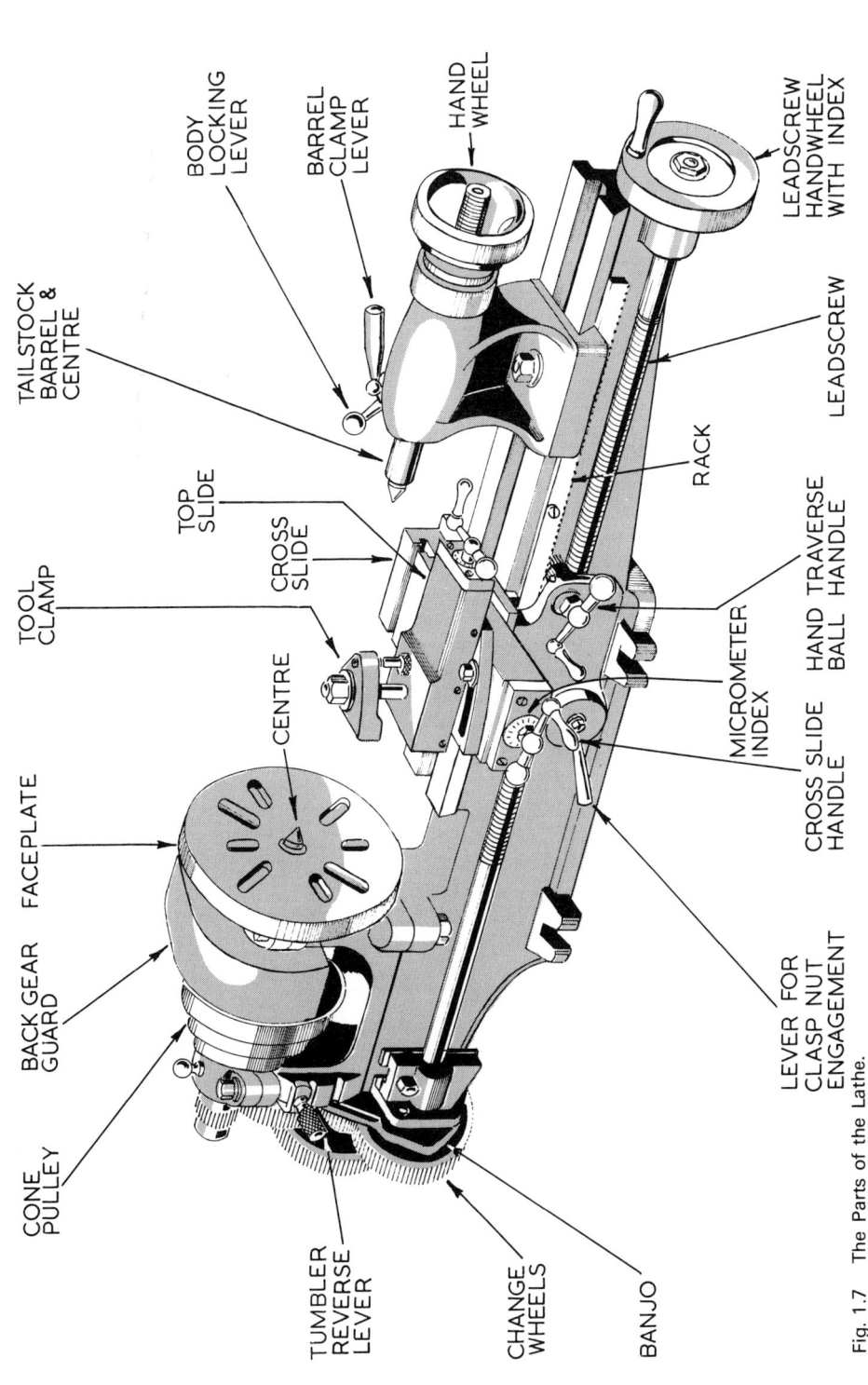

BODY LOCKING LEVER

BARREL CLAMP LEVER

HAND WHEEL

LEADSCREW HANDWHEEL WITH INDEX

TAILSTOCK BARREL & CENTRE

LEADSCREW

RACK

TOP SLIDE

CROSS SLIDE

TOOL CLAMP

CENTRE

MICROMETER INDEX

HAND TRAVERSE BALL HANDLE

CROSS SLIDE HANDLE

FACEPLATE

BACK GEAR GUARD

CONE PULLEY

LEVER FOR CLASP NUT ENGAGEMENT

TUMBLER REVERSE LEVER

CHANGE WHEELS

BANJO

Fig. 1.7   The Parts of the Lathe.

illustration, Fig. 1.7, depicts a typical small lathe showing the location of components that will be mentioned from time-to-time. If reference is made to this illustration it will help the reader to familiarise himself with the various parts of the lathe as they occur.

Amongst the designs left by that genious Leonardo da Vinci were those for a treadle lathe. As with many of his suggestions there is no certainty that this lathe was ever constructed. Had it been made it is possible that the lathe as a practical machine tool would have developed earlier than it did.

Be that as it may, the main impetus in development came from the clockmakers in the 18th century who needed machines that would produce the many intricate parts they required. In this way the lathe became a 'do-all' capable of not only turning but gear-cutting as well. At the same time the requirement for ornamental work led to the introduction of the rose engine and other ornamental turning lathes and devices.

Elementary Screw Cutting Machinery

The screw, because of its many applications, is probably the most important of the mechanical elements. It is said to have been invented by Ardeytas of Tarentum, a pythagorean philosopher and mathematician, about 400 B.C., but is generally associated with Archimedes who died in 212 B.C. Both Hero and Pliny have described wooden screws as having been applied to presses in the first century A.D., but do not appear to have left a description of how these screws were made.

On the other hand, Leonardo da Vinci has again left us with a machine design in which two leadscrews geared to the workshaft

Fig. 1.8   Screwcutting Lathe 1820

enabled the pitch of the screws being cut to be varied at will by changing the ratio between the driving and the driven gears. In this matter da Vinci's machine bears a strong resemblance to a device used in Woolwich Dockyard about 1850 and illustrated in Fig. 1.9. But in his design the leadscrews did not control individual toolposts as seen in the illustration, but acted together to advance a single toolpost spanning the whole machine.

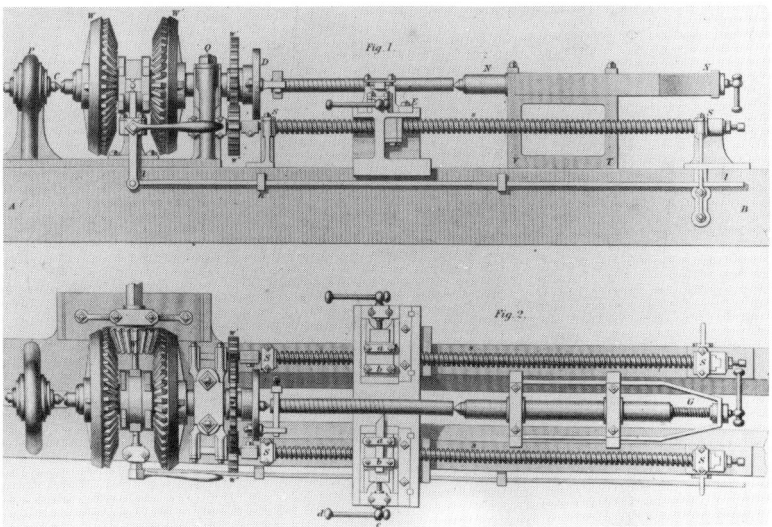

Fig. 1.9   Thread turning machine from Woolwich Dockyard 1850.

Leonardo da Vinci has recorded the use of the screw both in the making of other screws and the construction of machines, but whether or not his screw-making machines were ever made seems uncertain. Nevertheless, in this matter he was clearly capable of very original thought so, as he was demonstrably much in advance of his time, it seems that he owed little of this to his contemporaries.

The production of screw threads had evidently been exercising the minds of mechanics for a long while, but it was not till about 1700 that a practical solution to the problem presented itself. By this time the design of the lathe headstock had anvanced to the point that would allow the device to be put into practice.

The Traversing Mandrel

This device, introduced about 1700, consisted briefly in allowing the lathe mandrel to slide axially in its bearings under the control of a master screw attached to the end of the mandrel itself. When a fixed half nut was made to engage this screw the rotating mandrel moved forward on its axis. With a single point tool brought into contact with the work and held stationary a thread would be produced on it having the same pitch as that of the master screw attached to the mandrel.

A similar device was fitted to some Pittler capstan lathes, made presumably during the period of the Second World War 1939–1945,

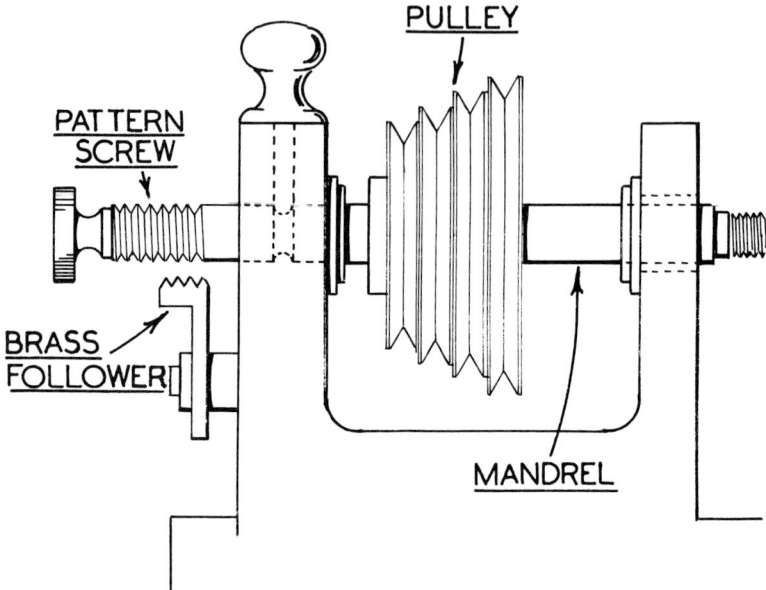

PULLEY

PATTERN
SCREW

BRASS
FOLLOWER

MANDREL

Fig. 1.10   The Traversing Mandrel.

so that components having short lengths of thread could be produced. Examples of these lathes were installed in a large works where the author was once employed, and there was a complete kit of equipment for use with them. But not much use was made of the traversing mandrel device, the operatives and planners preferring to use other methods with which they were more familiar for the batch production of screwed components. The Traversing Mandrel is depicted diagrammatically in Fig. 1.10.

The traversing mandrel was evidently well known to clockmakers and ornamental turners in the late 17th century, for the description of an ornamental lathe having the device at that time exists. This machine has two forms of tailstock, one apparently able to be set over so that tapers could be turned.

The limitations of the traversing mandrel soon became apparent, so it was not long before many first-class brains turned their attention to alternative methods of screw thread production. Thus the names of such eminent people as Henry Maudslay, David Wilkinson in America, Richard Roberts, James Nasmyth and by no means least, Joseph Whitworth, were concentrating on this problem and others connected with the improvement of the lathe as a practical machine tool.

For screwcutting purposes, the traversing mandrel was superseded by the leadscrew. This is a long screw running the full length of the lathe bed and usually in front of it, though some lathes locate them differently.

The leadscrew can be driven from the mandrel through a train of gears that may be changed at will. A nut attached to the saddle of the lathe engages the leadscrew, so, when the screw rotates the saddle will travel along the lathe bed. The ratio of the gearing that connects the mandrel and leadscrew together dictates the pitch of the screw that will be cut. For example if the mandrel itself turns at twice the speed of the leadscrew then the pitch of the thread cut on the work will be twice that of the leadscrew itself whatever that may be. Small lathes of the type already illustrated in Fig. 1.7 usually have leadscrews with a thread pitch of $\frac{1}{8}$in. so, when the mandrel makes two turns for each one of the leadscrew, the thread pitch cut will be $\frac{1}{16}$in.

Using the principle of which this example is the basis, it will be understood that, by selecting gears to connect the mandrel and leadscrew in the right ratio, a wide range of thread pitches can be cut in the lathe.

Henry Maudslay, 1771–1831, was employed originally by Bramah, the 'locksmith', but left his services in 1797 following a dispute over rates of pay. Bramah himself was a prolific inventor, numbering amongst his designs the hydraulic press and the beer engine, but there is little doubt that he owed much to Maudslay, as a practical mechanic, who perfected Bramah's locks.

Maudslay also produced a machine for winding springs and which set him to consider the problem of screw cutting.

The end of the century saw the production of two further lathes equipped to produce screw threads; one by a Frenchman named Senot, the other by David Wilkinson, an American, who introduced his machine in 1798.

It is at this point that the name of Richard Roberts 1789–1869 becomes important. He had originally been employed by Maudslay, from whom he must undoubtedly have absorbed some sound engineering practice, but in 1814 he left him and set up on his own in Manchester. His claim to fame rests on his introduction, in 1817, of the first practical back-geared headstock to be applied to a lathe.

This is a cardinal point in lathe development for it must be remembered that the Industrial Revolution was beginning to gather momentum so the size of components needed to be machined was consequently increasing rapidly. As a result the limitations of the existing plain lathes were only too apparent. The reader with practical experience of turning will not need to be reminded that the peripheral speed of the work has a direct bearing both on the life of the turning tool as well as the finish of the machined surface. Therefore, when a large casting, for example, is mounted on the faceplate the lathe mandrel itself must rotate at a relatively low speed in order to bring the peripheral speed of the work down to an acceptable figure.

Fig. 1.11   Maudslays Lathe 1810

Employing the flat, or round leather belts available at the time by themselves, it was really not possible to provide the slow speeds that were essential. If one explains that the transmitting power of these belts is directly proportional to their linear speed, it will be clear that at a low belt speed comparatively little power would be transmitted and consequently the turning of large objects difficult if not impossible. On the other hand if one was able to retain a high belt speed and, at the same time, get the lathe mandrel turning slowly then all the necessary conditions for heavy machining would be satisfied and the versatility of the lathe itself greatly increased.

Accordingly, as depicted in Fig. 1.12, Richard Roberts interposed a simple train of gears between the mandrel driving pulley and the mandrel itself.

In this way he doubled the range of speeds that could be provided. The mechanical arrangements of a machine so fitted are simple. The driving pulley, instead of being secured directly to the mandrel is now free to turn on it and has a pinion attached to its outer end. This pinion engages the large wheel of the back gear cluster, the small pinion of this unit being in gear with the bull wheel which is affixed to the mandrel itself.

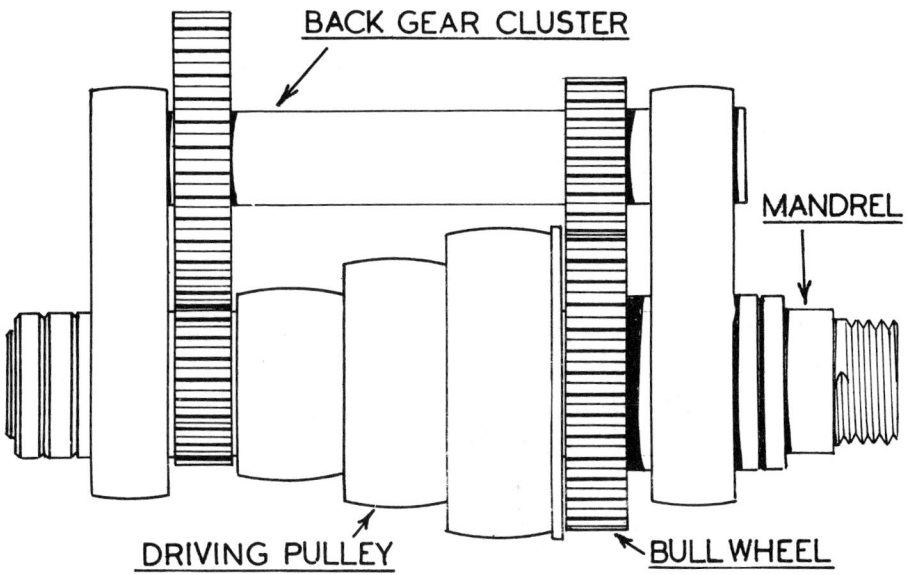

Fig. 1.12   Back Gear.

Accordingly, depending upon the number of steps on the driving pulley, a range of low mandrel speeds is now available. When, on the other hand, a higher mandrel speed is needed, the back gear cluster is thrown out of engagement and the driving pulley is connected directly with the bull wheel by means of a sliding pin or some other device. A typical lay-out for a back gear system as applied to a modern inexpensive but high-class small lathe is illustrated in Fig. 1.13 where it will be seen that the gear cluster itself is a close-coupled unit that may be quickly thrown in or out of engagement and secured by the unit seen on the end of the back gear spindle.

Fig. 1.13   Back Gear for modern small lathe.

As an example of the range of speeds available with the device the following may be regarded as representative:

Direct Drive			With back gear engaged		
840	490	280	145	85	48
	r.p.m.			r.p.m.	

Roberts was evidently an enterprising tradesman well able to gauge the market for machine tools and their equipment that was fast developing, for by 1821 he was advertising himself as a maker of division plates and worm gears, some of these last being of large size. Roberts had been an employee of Wilkinson 1725–1805, who invented the boring machine used by James Watt in his improvements to the steam engine, but there is no suggestion that Wilkinson himself provided any significant contribution towards the development of the lathe itself.

Perhaps the most important name that comes to mind in connection with the manufacture of machines during the 19th century is that of Joseph Whitworth 1805–1870. He was another product of the famous Maudslay organisation but does not appear to have stayed with it for very long; long enough, however, to have absorbed much of the technical know-how for which the firm was famous. So in 1833 we find him in business for himself as a machine tool manufacturer; the first of them in fact, because apart from one or two manufacturers of tools for clockmaking, all the other makers were producing machines to be used in their own works and so not for re-sale.

Whitworth exhibited on a large scale at the 1851 Exhibition in Hyde Park, in fact the range of machines offered by him far surpassed that of any other exhibitor.

Whitworth made many contributions to the development of machine tools generally. However, in connection with the lathe in particular, perhaps his most important work was the patenting in 1839 of the claspnut used in connection with the leadscrew.

Up to this time any lathe fitted with a leadscrew had a fixed nut attached to the saddle that could not be disengaged at will. Consequently when the saddle or carriage driven from the leadscrew had come to the end of its travel it could not be returned quickly to the beginning again but had laboriously to be wound back by turning the leadscrew itself, or perhaps running the lathe in reverse.

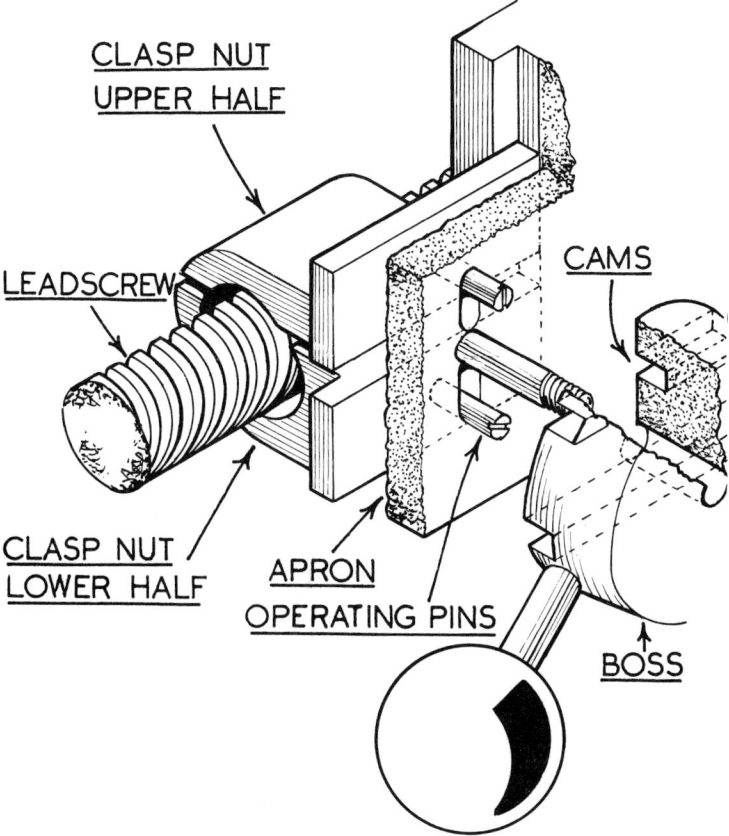

Fig. 1.14   The Clasp Nut.

Clearly this was an enormously time-wasting proceeding that must have added greatly to the expense of all general machining operations.

Joseph Whitworth's patent provided for the dividing of the nut axially and the mounting of each half in a slideway forming part of the lathe saddle itself, the opening and closing of the nut being controlled by a lever mounted on the front of the saddle apron.

At the same a pinion or a train of gears, mounted behind the apron and engaging a rack fastened to the front of the lathe, allowed the operative to wind the saddle back quickly as soon as the clasp nut was opened.

In this way the time lost between cuts along the work was reduced to a fraction of what it had been previously.

The clasp nut and its disposition are illustrated in Fig. 1.14.

Turning Tools

We come now to that most important subject that of the tools used for machining the work and the methods for mounting them in the lathe.

For centuries all turning was carried out with hand tools of the type illustrated in Fig. 1.15 for the most part on wood.

This illustration, taken from 'A Panorama of Science and Art', depicts the battery of hand tools available to the turner at the time of

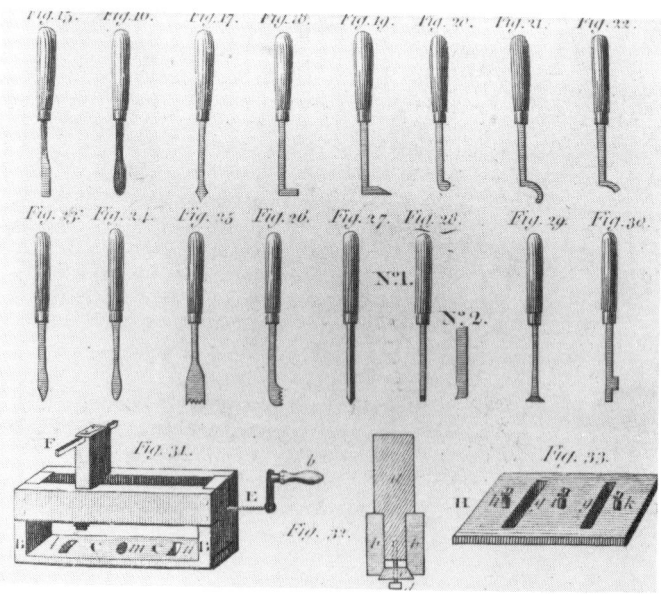

Fig. 1.15   Hand Turning Tools.

the book's publication about 1750. Many of these have not survived the passage of time but four have. These are the tools marked 'Fig. 25', 'Fig. 26', 'Fig. 27' and 'Fig. 28'. The first two are used in the cutting of threads whilst the others are employed firstly to shape the part being turned and then to part it from the parent material.

Fig. 1.16   Hand Turning Tools in use.

The tools were supported on a hand rest of the pattern depicted in Fig. 1.17 and bearing a strong resemblance to the rests sometimes available today. As will be seen the rest proper, consisting of the bar member indicated by the latter 'M' is supported in the boss of a forked casting that enables the bar to be set at the correct height and then clamped on the bed of the lathe as close to the work as may be required.

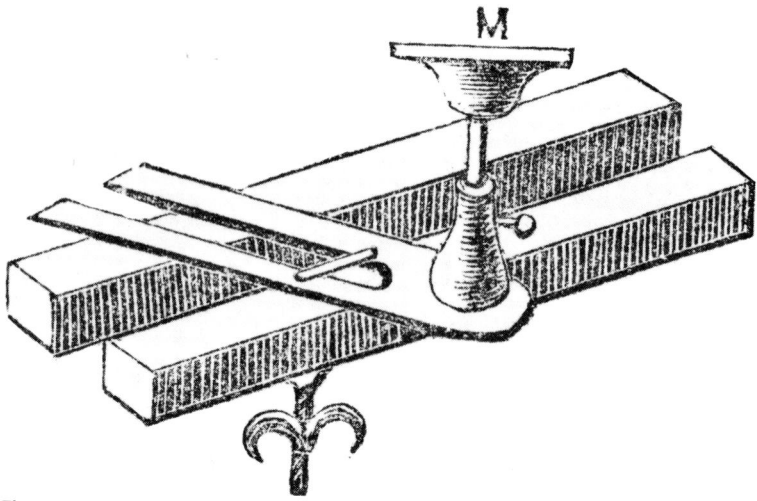

Fig. 1.17   The Hand Rest.

For the turning of metal the principle tool was the graver. This tool, which is still used by the author and many other operators for a number of machining duties, it made from a length of tool steel shaped as illustrated in Fig. 1.18. A lozenge shaped facet is ground at the point and the tool as a whole is set in a wooden handle so that the operator can use it. Space does not allow a detailed description of the manner in which the tool is applied, nor is this the book in which this should be described.

The author has dealt fully with this matter elsewhere in a book that is virtually a beginners guide to the workshop. 'The Novice's Workshop'—Model & Allied Press Limited. One of the advantages when using the graver is that the hand rest can be placed so close to the work. Consequently the point of the tool does not overhang, so there is little fear of it 'digging in'. Unfortunately this advantage does not

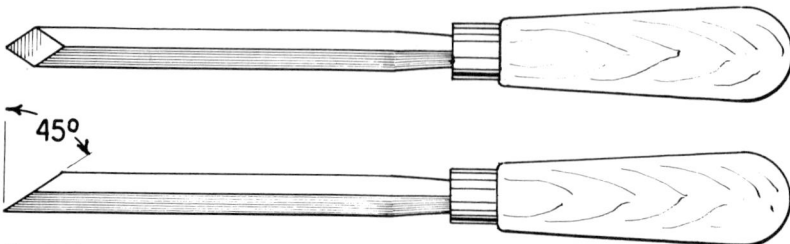

Fig. 1.18   The Hand Graver.

obtain if a boring tool supported by the hand rest is in use. Though one has never tried the experiment on a component of large size one can readily imagine the difficulties and appreciate the reason for the very long handles with which they were provided.

Writing, in about 1800, one authority on the subject has described the work itself as being mounted on the mandrel with its outer end supported in a species of rest consisting of a series of holes of a size sufficient to accommodate the outside diameter of a range of work and able to be brought to bear on the particular piece of work in hand.

'Then,', says our authority, 'the end of the cylinder to be bored, being placed in the hole which fits it, the boring tool is held upon a rest against its centre and the boring may be then performed with great accuracy to any required depth.' But to quote another authority: 'It all depends upon what you mean by accuracy.'

Certainly, the steam engine makers were not getting either the accuracy nor the finish they needed, for the standard of workmanship resulting from hand methods did not allow it.

On Watt's steam engine the errors in the machining and in the fits of cylinders and pistons were very great. To quote the ninth edition of the Encyclopaedia Britannica, 'James Watt in 1769 was fain to be content with a cylinder for his Fire-engine of which, though it was but 18 in. in bore, the diameter in one place exceeded that in another by about $\frac{3}{8}$ in., its piston was not unnaturally leaky though he packed it with paper, cork, putty, pasteboard and an old hat.'

With only such a poor standard of machining being available it was clear that something had to be done to rectify matters. The ensuing development took the form of rests of one type or another in which the tool itself was firmly clamped but could be moved across or along the work taking a cut of predetermined depth.

Earl Stanhope, who had perfected a printing process, appears to have been one of the first in England to make use of an elementary form of slide rest. In France a slide rest similar to that illustrated in Fig. 1.19 representing the types in use for ornamental turning. Henry Maudslay, of whom we have already heard, seems to have set a similar device to work in Bramah's workshop about 1794.

From the outset it was realised that, for all practical purposes, the rest should be able to operate in two planes at right angles to one another. This, in effect, involved the use of two separate slides, each controlled by its own feed screw, one superimposed on the other, the upper slide carrying the tool post and also being capable of being swung at an angle to the work so that tapers could be turned on it.

The first attempts at making the lathe self-acting, that is by moving the slides automatically, centred on moving the top slide only. An example, an exhibit from the 1851 Crystal Palace Exhibition, is

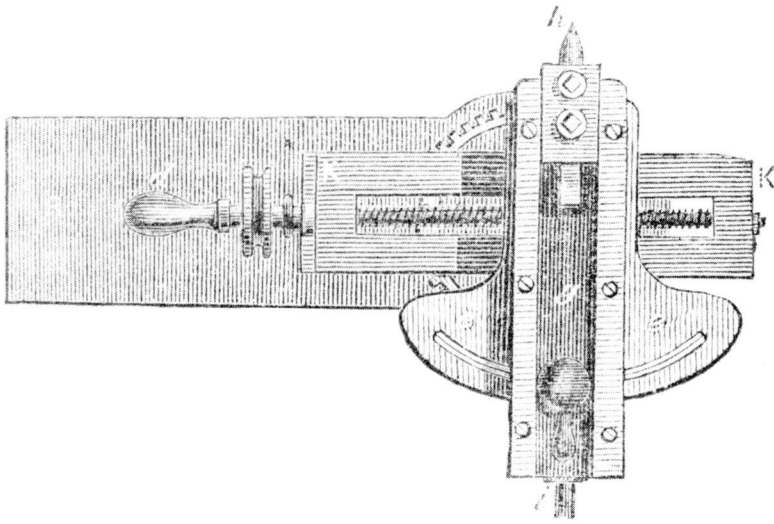

Fig. 1.19   View of a slide rest for ornamental turning.

depicted in Fig. 1.20. Here, a striker affixed to the work, is made to engage a star wheel forming an extension element of the top-slide feed screw, the whole set-up an undoubtedly noisy contrivance, reminding one irresistibly of the cyclometer drive once fitted to bicycles.

A more sophisticated method of carrying out the same idea is illustrated in Fig. 1.21. In this case, which depicts the late George

Fig. 1.20   A self-acting slide rest from the 1851 Exhibition.

Fig. 1.21   The Gearge Adams precision lathe.

Adams' top-slide self-act as fitted to his $2\frac{1}{2}$m centre precision lathe of some 30 years ago, the feed screw itself is extended to carry a telescopic shaft fitted with a dog-clutch for rapid engagement and disengagement, this assembly being driven from the mandrel.

It will be appreciated that no shake can be tolerated in the slides themselves, for if there is any then the work surface will be machined inaccurately as well as roughly. To enable all shake to be removed, as well as to ensure that they move smoothly, the slides are fitted with what are known as gib strips placed in the location shown in Fig. 1.22.

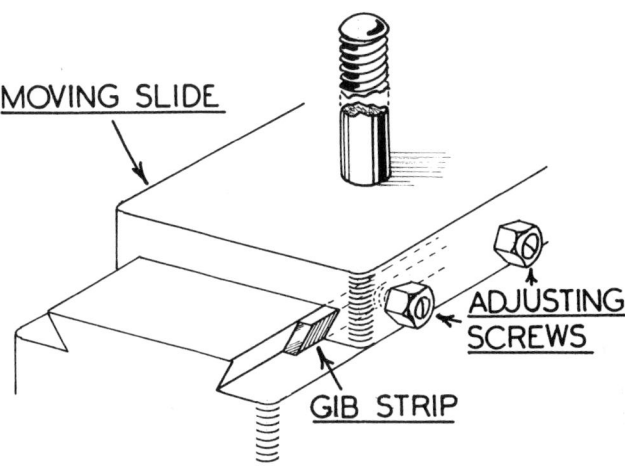

MOVING SLIDE

ADJUSTING
SCREWS

GIB STRIP

Fig. 1.22   The GIB Strip.

For the sake of clarity the gib strip is shown extended from its normal position flush with the end of the moving slide. The strip is held in place by the adjusting screws making contact with dimples formed in the strip itself.

Automatic Cross Feed

We have already seen how the leadscrew, driven from the lathe mandrel, is used to traverse the saddle along the lathe bed so that a uniform rate of tool feed can be provided, so it was not long therefore before the logical step of mechanically driving the cross-slide feed screw was put into practice. The illustration, Fig. 1.23, shows a large lathe of the type made about 1880 and demonstrates the arrangement of the train of gears needed to provide an automatic cross-feed. The train was driven from a worm gear keyed to the leadscrew and able to travel along it, a keyway running axially along its length causing the driving gear to rotate.

Fig. 1.23   Large Lathe (about 1880) showing gears for auto cross feed.

Later it was found that the continual use of the leadscrew itself as a means of driving the saddle along the bed, for purposes other than the cutting of threads, was detrimental to its accuracy because wear tended to be concentrated in an area close to the headstock where most of the turning work took place.

As we shall see presently, later developments, particularly as applied to larger types of machine, lead to the leadscrew being superseded for all purposes other than screwcutting.

But to return to Joseph Whitworth. The illustration, Fig. 1.24, depicts in section his couble slide rest. This device incorporated two independent compound slide rests that could be used for adjusting the depth of cut to be applied to the work; in addition the saddle cross-feed,

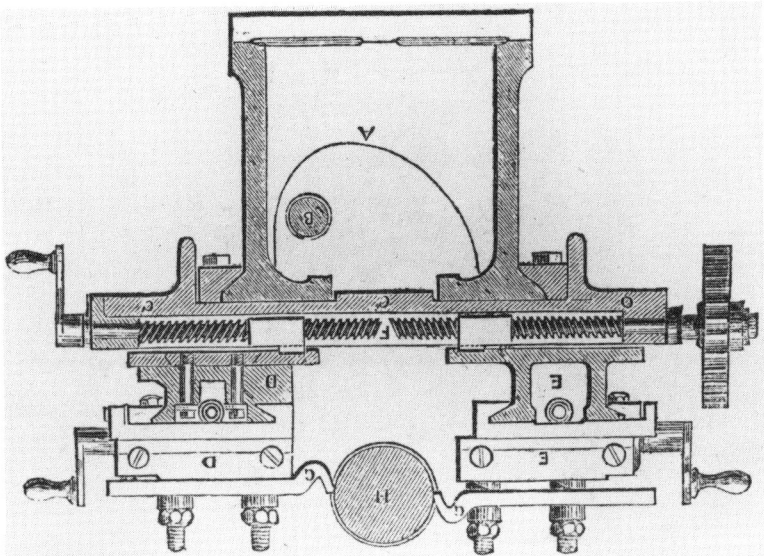

Fig. 1.24   Whitworths Double Slide Rest.

as well as being capable of operation manually, could be power driven through the gear wheel seen attached to the feed screw.

By 1840 Whitworth was supplying lathes fitted with the basic refinements, such as automatic sliding and surfacing, that we now take for granted, so it is hardly surprising to learn as we have done that his exhibits at the Crystal Palace far exceeded those of any other maker both in quantity or quality.

CHAPTER TWO

THE MODERN CENTRE LATHE

BEFORE GOING ON to discuss the modern lathe we must take into consideration the developments in the means of driving machine tools generally, for the developments in this field have had a great bearing on the design of the tools themselves.

When it became clear that the rapidly increasing requirements of industry were not going to be met by the somewhat primitive arrangements described earlier, engineers set about harnessing water power and then the steam engine to the driving of machine tools. For the most part the prime mover was coupled by belting to a shaft running the length of the workshop. From this lineshaft, as it was called counter-shafts were driven, again by flat belting, so that each individual machine could be connected to or disconnected from the main drive at will.

As will be seen in the illustration, Fig. 2.1, the countershaft was provided with a pulley for driving the particular machine tool and a pair of pulleys, one secured to the countershaft, the other running loosely on it. The belt from the lineshaft could be moved across from one pulley to the other thus being connected or disconnected from the main drive as required.

As may be imagined in a large workshop with many machines, this arrangement involved the employment of a perfect forest of belting, that had to be supervised by a considerable number of very experienced men in order to service adequately the belting and the transmission equipment driven by it. Nevertheless, when properly maintained, the lineshaft and countershaft system was quite satisfactory. But, on the large scale, the belting was somewhat obstructive of light. However, a development in the lathe did much to remove this obstruction by the elimination of the countershaft itself.

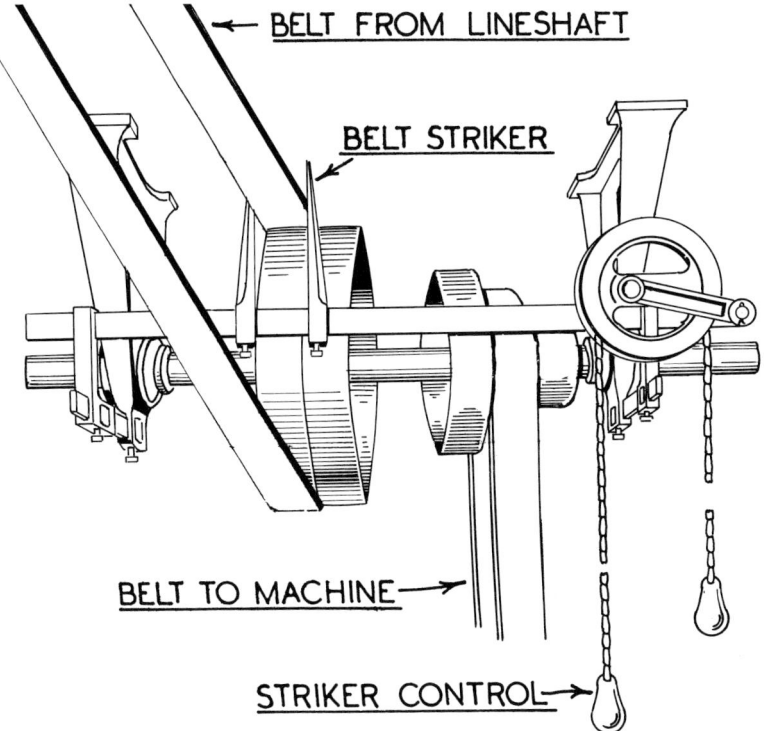

← BELT FROM LINESHAFT

BELT STRIKER

BELT TO MACHINE →

STRIKER CONTROL →

Fig. 2.1   The Lineshaft and Countershaft system.

The All-Geared Headstock

Up till about 1900 the drive to the lathe headstock and the variation of speed obtained at the mandrel nose was derived from a stepped cone pulley and back gear as has already been described. An all-geared head was then introduced from America enabling the operative to change speeds rapidly, instead of having to throw the belt from one step of the countershaft pulley to another in order to secure a change of speed. This throwing of the belt inevitably involved some waste of time, especially in works where the drives to the machines were of some length, and a special pole had to be used for the purpose. Many operatives, however, were adept at 'throwing' moving belts by hand, probably in defiance of safety regulations though in one works where the author served part of his time no one seemed to think this practice unusual.

The gearbox fitted to an all-geared head was very similar in conception to one used in a motor car. Indeed the idea of applying a gearbox to the lathe may have come from this source, see Fig. 2.2. In addition to the gearbox the lathe was sometimes provided with a clutch through which the drive from the lineshaft was taken. A single pulley only was then needed directly connected to a corresponding pulley mounted on the lineshaft, thus the countershaft was eliminated so removing some of the obstruction to light caused by the belting employed with it.

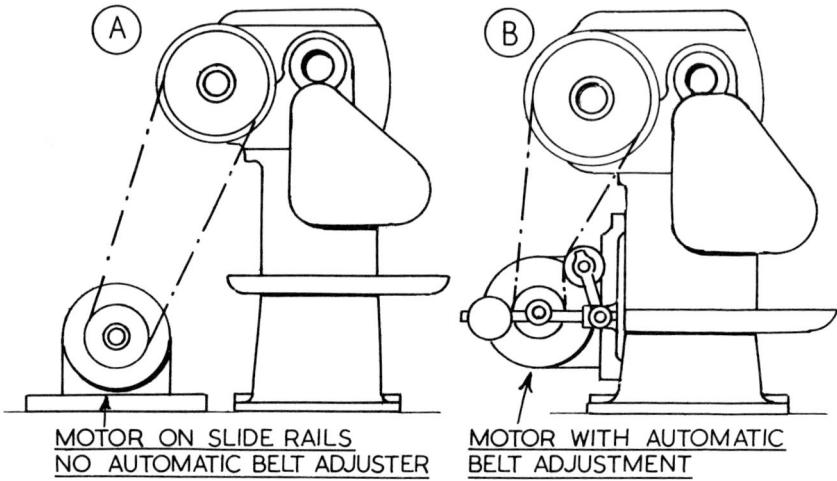

MOTOR ON SLIDE RAILS
NO AUTOMATIC BELT ADJUSTER

MOTOR WITH AUTOMATIC
BELT ADJUSTMENT

Fig. 2.2   All-geared headstock (with V-Rope Drive).

With the advent of electrical supplies, the water turbine, the steam engine and later the gas engine were replaced by electric motors, usually of some size, placed overhead adjacent to the lineshafting which was then split up into divisions enabling whole sections of the machine shop to be shut down when needed.

The next step in the modernising of the driving arrangements was to do away with lineshafting altogether and replace it with individual electric motor drives to each machine. At first flat belts continued to be used, but these were not too successful because of the shortness of the distance between the centre of the driving pulleys and those of the machine tools themselves. Various devices were tried in order to improve the belt grip, but the equipment was somewhat cumbersome, so it was not until the introduction of the V-rope that a satisfactory short-centre drive was finally attained.

The V-Rope

Until about 1914, when a chain-cum-belt drive was beginning to be used, motorcycles had used belts of 'V' section with success despite the adverse conditions under which they had to run, so it is a little surprising that these belts did not find their way into the workshop earlier. The greatly increased gripping power of the V-rope enabled extremely short-centred drives to be designed, thus providing the lathe manufacturers with an opportunity of setting out a really compact self-contained electric drive.

The V-rope is made as an endless belt of varying lengths and sections and can be used either singly or in multiples. Employed in the last way a drive of great flexibility combined with the maximum of grip is obtainable. This is the arrangement usually employed in modern all-geared lathe headstocks having a single pulley drive as illustrated in Fig. 2.2. At 'A' an arrangement commonly employed is illustrated. Here the electric motor, mounted on slide rails is bolted to the floor. Adjustment of the driving belt is effected by sliding the motor along the rails and securing it when correct belt tension has been achieved.

An alternative method of mounting the motor, and one that reduces the floor space occupied, is depicted at 'B'. The motor is bolted directly to the lathe, a pair of rails usually being provided to carry it. Belt tension can be set roughly by positioning the motor correctly on the rails. The example shown has, additionally, an automatic tensioning device consisting of a counterpoise carrying a pulley that bears on the back of the belt itself.

The V-rope has also revolutionised the drive of many small lathes, particularly those used in the amateur field, one of which is seen in the illustration, Fig. 2.3. For heavy lathes silent chains have sometimes been used to provide the drive from an electric motor, but for the most part multiple V-ropes have taken their place.

A further development in the arrangements for a self-contained or built-in electric motor drive is illustrated in Fig. 2.4. Here, as will be seen, the belt is dispensed with altogether. Instead a flange-mounted electric motor is bolted directly to the casting of the lathe at its headstock end and the drive is taken, through a clutch, directly to the gearbox itself. In a heavy lathe this is undoubtedly the only way to arrange matters when really positive non-slip power transmission is essential.

Modern Headstock Bearing Arrangements

The bearings of early lathes were of plain type with the mandrel running directly in the headstock or in bushes set in the casting.

Fig. 2.3   Modern Light Lathe with V-Rope Drive.

This arrangement still obtains in many light lathes today and two examples are illustrated in fig. 2.5 where the addition of a ball race to absorb the end thrust may be seen. For the most part instrument lathes and small lathes in the precision class like those illustrated in Fig. 2.6 make use of such an arrangement but without the ball race.

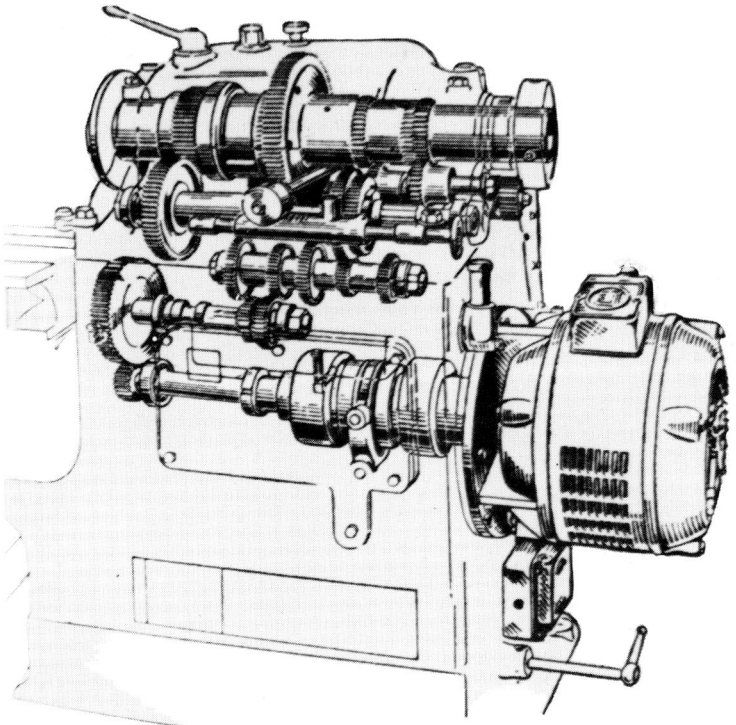

Fig. 2.4   Built-in Electric Motor with shaft drive.

Instead a double taper on the mandrel nose, as illustrated in Fig. 2.7 provides the ideal thrust absorbing device in tools intended for fine turning to close limits. The mandrels themselves are hardened, and for the most part, are supported in a hardened steel bearing set in the casting though some machines employ the layout depicted in the illustration.

For the high speeds required in a modern production lathe plain mandrel bearings are quite inacceptable. Various combinations of ball and roller races are therefore employed to ensure that high

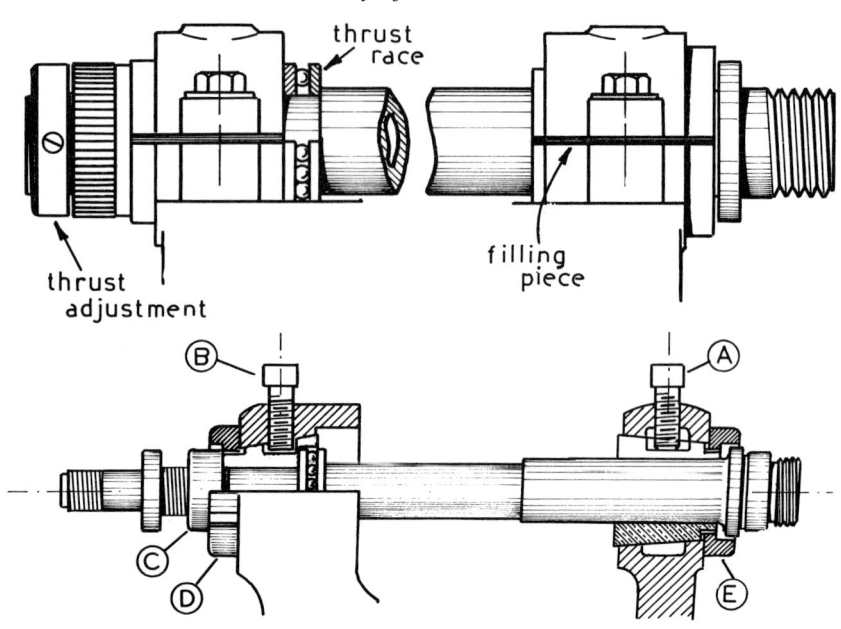

Fig. 2.5   Types of Headstock Plain Bearing arrangement.

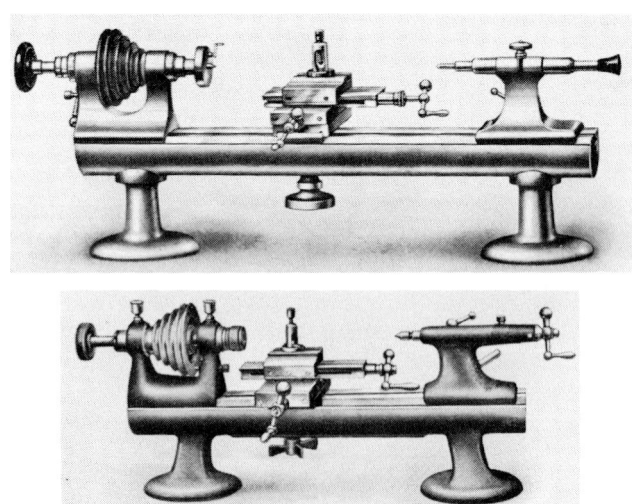

Fig. 2.6   Light Lathes in the precision class.

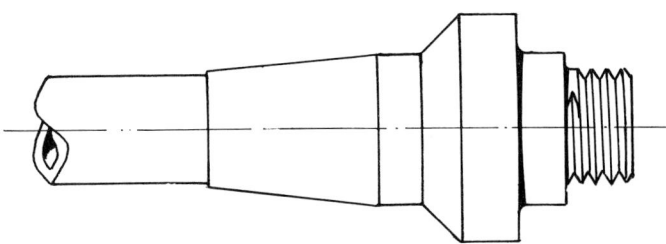

Fig. 2.7   Lathe mandrel with double taper.

speeds can be maintained. Apart from this advantage ball or roller bearings are far simpler to lubricate, in addition their maintenance and adjustment is easier to carry out.

An example of the use of roller bearings in a lathe headstock is illustrated in Fig. 2.8.

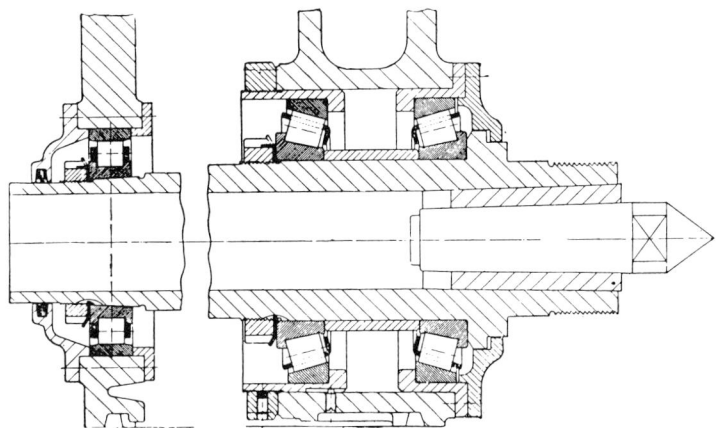

Fig. 2.8   Arrangement of Roller Bearings in a Lathe headstock.

The Norton Gear Box

When much screw threading had to be carried out, or rapid changes in feed speeds needed to be made, it was manifestly impossible to remain content with the change wheel system by itself. Setting up the train of wheels was always a slow operation and industry was on the look-out for some system by which the pitch of the screw to be cut or the rate of tool feed could be selected as easily as the spindle speed itself. The answer to the problem was the Hendy Norton gearbox illustrated in Fig. 2.9, where the mechanism is displayed.

GEAR CHANGING BY HAND-LEVER

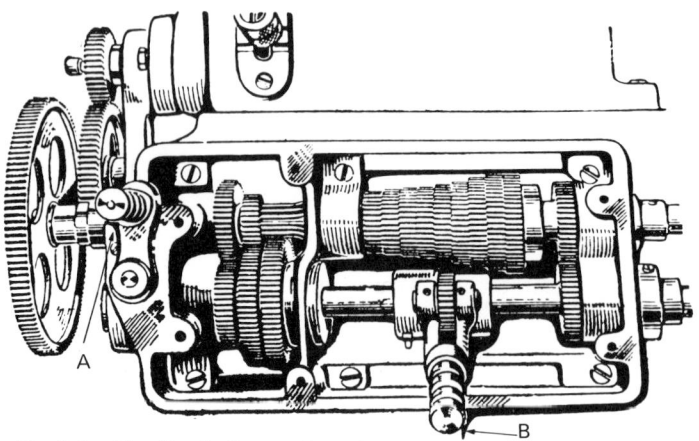

Fig. 2.9   The Hendy-Norton Gear Box.

Fig. 2.10   The Norton Gear Box fitted to a Myford Lathe.

Two shafts project from the right-hand side of the gearbox. The upper shaft is the leadscrew, used for screw cutting only, whilst the lower is the feed shaft that is the main drive to both the saddle and cross-slide enabling both sliding and surfacing cuts to be made automatically so saving the operative having to hand-feed them. Two levers are needed to effect a change of pitch or of feed. The lever on the left marked 'A' will select three different ratios whilst the lever 'B' carrying the tumbler pinion can be moved to bring it into mesh with one of the nine gears in the cone seen at the upper area of the gearbox. In this way 27 variants of speed can be made.

Fig. 2.11   A large modern Lathe.

An example of a Norton gearbox fitted to a modern light lathe is illustrated in Fig. 2.10. The device was developed in America about the same time as the all-geared headstock, that is in 1892, and was quickly adopted by many lathe manufacturers.

CHAPTER THREE

LATHE TOOLS

SOME OF THE TOOLS for use with the hand turning rest have already
been described so we need not concern ourselves further with them
here.

Initially lathe tools were forged from rectangular steel bar capable
of being hardened and ground to produce a durable cutting edge able to
withstand the wear imposed by the machining processes. The shapes
needed were few, consisting in the main of tools necessary for stock
removal both inside as well as outside the work, with the addition of a
pair of external and internal screwcutting tools and one tool to be used
for parting work from bar material. A representative set of tools is
illustrated in Fig. 3.1. The material used was cast steel, easily forged in
the smithy, hardened by heating to red heat and dipping in cold water, a
simple tempering process being applied subsequently to fit them for
use.

Tools of this nature were mounted on the top-slide and held down
by a simple clamp, packing being placed under the shank to bring
the cutting edge to the correct height.

Originally the tools were held down, for the most part, by steel
bars passing over studs set in the lathe top-slide, a practice that may
still find favour in many large lathes today. But later the two forms
of toolpost illustrated in Figs. 3.2 and 3.3 were introduced. In the
first of these the tool is held down on packing when necessary by a
pair of screws set in a cast-iron or steel block, itself secured to the
top slide by a stud and nut and able to be swung around, if needed,
the more easily to align the turning tool with the work.

In the second type illustrated the clamp comprises an iron casting,
fitted with a levelling screw, securing the tool by a single stud and nut,
the latter bearing on a spherical washer to take care of any misalign-
ment of the clamp.

The American type of toolpost was introduced towards the end
of the 19th century. This toolpost is illustrated in Fig. 3.4. The tool
itself is mounted on a 'boat' which is curved on its undeside and is

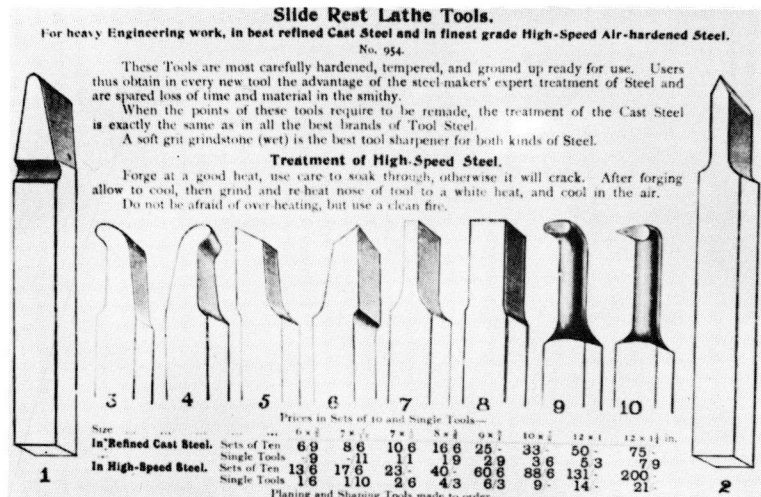

Slide Rest Lathe Tools.
For heavy Engineering work, in best refined Cast Steel and in finest grade High-Speed Air-hardened Steel.

No. 954.

These Tools are most carefully hardened, tempered, and ground up ready for use. Users thus obtain in every new tool the advantage of the steel-makers' expert treatment of Steel and are spared loss of time and material in the smithy.

When the points of these tools require to be remade, the treatment of the Cast Steel is exactly the same as in all the best brands of Tool Steel.

A soft grit grindstone (wet) is the best tool sharpener for both kinds of Steel.

Treatment of High-Speed Steel.

Forge at a good heat, use care to soak through, otherwise it will crack. After forging allow to cool, then grind and re-heat nose of tool to a white heat, and cool in the air. Do not be afraid of over-heating, but use a clean fire.

Prices in Sets of 10 and Single Tools—

Size	...	...	...	6 × ⅜	7 × ⁷⁄₁₆	7 × ½	8 × ⅝	9 × ¾	10 × ⅞	12 × 1	12 × 1¼ in.
In Refined Cast Steel.	Sets of Ten	6 9	8 6	10 6	16 6	25 -	33 -	50 -	75 -
	Single Tools	9	11	11	19	29	36	53	7 9
In High-Speed Steel.	Sets of Ten	13 6	17 6	23 -	40 -	60 6	88 6	131 -	200 -
	Single Tools	1 6	1 10	2 6	4 3	6 3	9 -	14 -	21 -

Planing and Shaping Tools made to order.

Fig. 3.1 A representative collection of Lathe Tools.

seated on a spherical washer. In this way the tool may be rocked to bring cutting edge to the correct height for turning and then locked by the screw at the top of the toolpost.

As will be seen the whole assemblage fits into a T-slot machined in the top-slide itself. It has the advantage that the tool can be quickly swung around and adjusted to bear on the work. But it suffers from the disadvantage of needing special offset tool holders when machining close to the headstock. In addition the particular method of setting the tool at centre height may, under certain circumstances, upset the rake and clearance angles of the tool as it approaches the work.

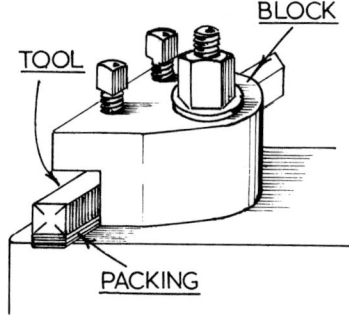

Fig. 3.2 Toolpost.

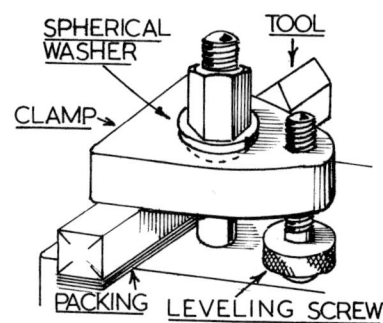

Fig. 3.3 Toolpost.

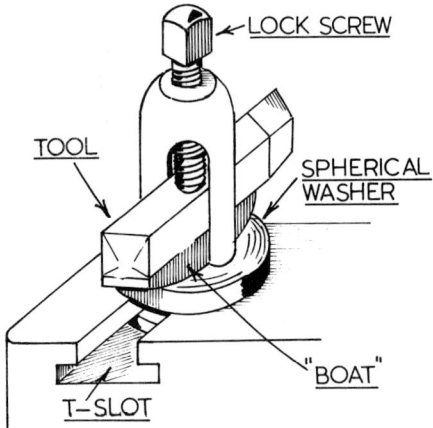

Fig. 3.4   The American Toolpost.

A modern application of the 'boat' as used in setting a lathe tool to the correct centre height is illustrated in Fig. 3.5. This is the form of lathe tool patented by the Myford Engineering Company for use on a top-slide provided with the standard type of English toolpost. The 'boat' has a convex or raised curved surface registering with a corresponding curved seating machined in the shank of the tool

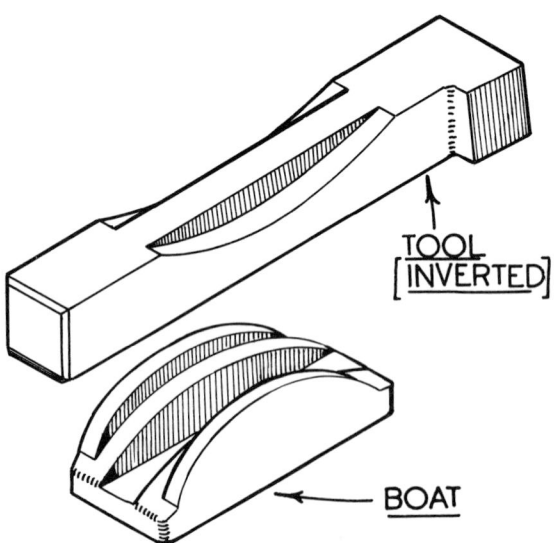

Fig. 3.5   Myford Tool and 'Boat'.

itself. In this way the tool point can be quickly adjusted with the same facility as with the American toolpost.

One further toolpost for single tools deserves mention. This is the post once fitted to the light lathe made by Drummond Brothers of Guildford. Our illustration shows the device in a form modified for use directly on the cross-slide though, in its production form of course, this toolpost formed part of the lathe top-slide.

Fig. 3.6   Drummond Toolpost.

As will be seen in Fig. 3.6 the tool carriage is a steel block having a seating into which the tool is passed and secured by a pair of screws. The steel block itself is split and is made a sliding fit over the upright member to which it can be clamped by the cross-bolt seen in the illustration. In this way the point of the tool can be quickly set at the correct height without any packing being used. In production top-slides the vertical member formed part of the iron casting from which the slide was machined.

The Capstan Head or 4-way Toolpost

It will be appreciated that with toolposts designed to accept a single tool only, much time can be occupied in changing over from one tool to another. For commercial purposes, and even when the saving of time is not of much financial importance, the waste of time taken in the change over may not be acceptable. The solution is to make use of the capstan head toolpost illustrated in Fig. 3.7. Here, four tools at a time

Fig. 3.7   Capstan Head Toolpost.

can be mounted and brought into play as required. When a number of identical parts have to be machined it is, of course, essential that, when the capstan head toolpost is rotated and the tools themselves are brought into play successively they shall always individually be clamped at exactly the same point as previously, otherwise none of the parts will be machined to the required dimensional limits.

The capstan head toolpost is often used in conjunction with a toolpost set at the back of the lathe cross-slide. A parting tool is usually mounted here, its purpose being to sever the part being made from the bar material caught in the chuck, Fig. 3.8.

Fig. 3·8 Back Toolpost.

Modern Turning Tools

The rapid rise in the rate components were required to be machined at the latter half of the 19th century, as well as the increase in the toughness of the material from which they were being made, quickly rendered tools made from forged steel obsolete.

The speed at which turning had to be performed rapidly blunted these tools which had, in consequence, to be sharpened repeatedly. Manufacturers therefore sought the assistance of the steel makers in an effort to overcome the trouble and were given a new steel now commonly called 'high-speed steel'. With this material workshop processes, involving cutting tools of all types, were substantially accellerated.

Naturally a high grade tool steel of this nature is a somewhat expensive commodity so steps need to be taken to ensure that it is used as economically as possible. Where large lathe tools are involved it is usual to weld the high-speed steel head on to a shank made from high-tensile steel that cannot itself be hardened but has the necessary strength to withstand the loads imposed by turning operations. Such a tool is illustrated in Fig. 3.9.

High-speed steel tools are also available in the form of bits in short rectangular section pieces, ground all over and hardened and tempered ready for the operative to shape as required.

These bits are suitable for mounting directly in the lathe toolpost, but, in the smaller sizes, are most conveniently secured in the type of toolholder illustrated in Fig. 3.9 at 'B'. Here a shank of good quality high-tensile steel has a seating for the tool bit, sometimes at an angle as shown, machined in the shank. The tool can be adjusted for stand-out and then locked by the screw seen projecting from the bars on the upper side of the shank.

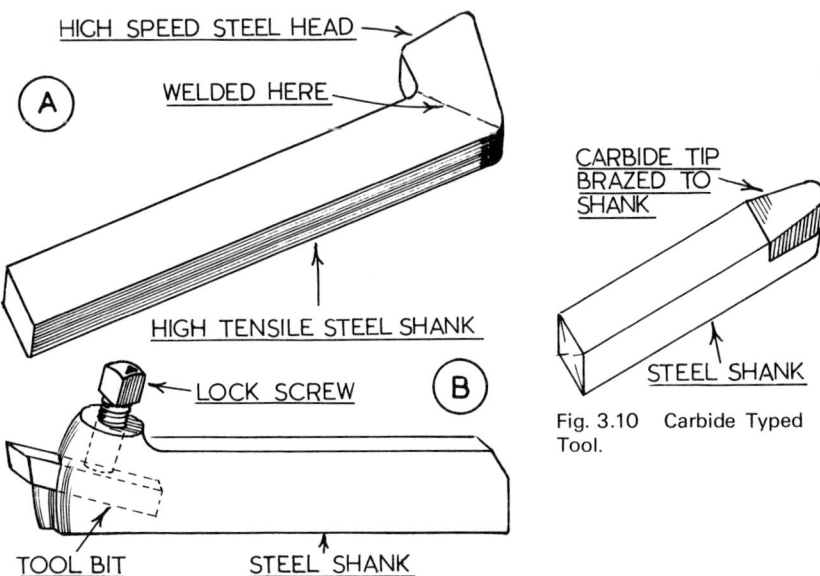

Fig. 3.10   Carbide Typed Tool.

Fig. 3·9   Methods of mounting High-Speed Steel Tools.

Whilst tools made from high-speed steel have retained their popularity for general machining purposes, the requirements of industry for materials that would cut steels of ever-increasing toughness and hardness led to the introduction of such compounds as tungsten carbide and 'Stellite'. The former is a sintered product in which the grains of the material are forced together under great pressure to form a composite tool tip, of no great mechanical strength in itself, which can be secured to a shank of high-tensile steel by a brazing process. A typical tool is illustrated in Fig. 3.10.

One further material has for many years been made into lathe tools. This is 'Stellite', a composite material made by the Deloro

Stellite Company. It is formed into complete tool bits of varying sizes and can be used directly without the aid of a toolholder. A characteristic of 'stellite' is that the hotter it gets as a result of machining operations, the tougher it becomes.

Experiments have also been conducted into the use of ceramics as metal-cutting substances. Of these probably the best known is 'Syntox', a material previously used as an electrical insulator but, when formed into a suitable shape and mounted in a holder designed for the purpose, well able to withstand heavy cutting at high speed.

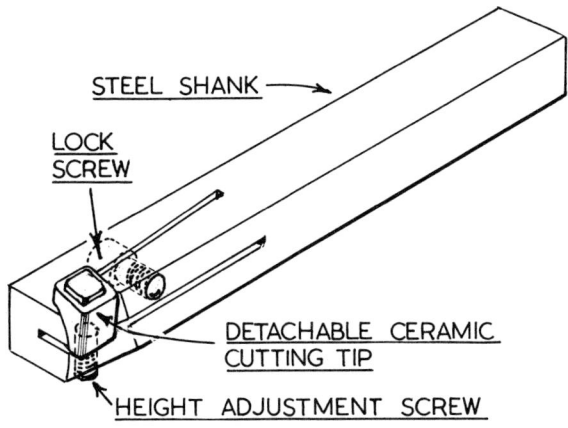

Fig. 3.11   Ceramic Tipped Tool.

These 'Syntox' points are four-sided so that when one face has become blunted another can be presented to the work. When all the four sides have been rendered unserviceable, the point is thrown away and a new one mounted.

A typical example of a ceramic-tipped tool is illustrated in Fig. 3.11. This is the product of the English Steel Corporation. The tool has a steel shank some 6in. long and $\frac{3}{4}$in. square into which the ceramic tip is set. A clamp screw holds the tip in place whilst a second screw is provided to adjust the height of its cutting edge. Front clearance for the tip is effected by canting it towards the work at an angle of 15°.

An interesting provision is the chip breaker formed on the upper surface of the tip itself. This comprises a projection about 0·030in. high against which the chip abuts as it comes off the work, thus breaking up the chip and reducing it to manageable proportions.

One of the more important tools that the turner needs to use is the parting tool. With it he is able to machine grooves in selected

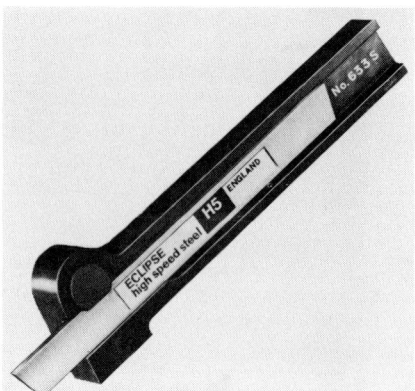

Fig. 3.12   Eclipse Parting Tool and Holder.

components or to sever parts from parent material after he has finished machining them.

In the past parting tools were forged from cast steel but latterly, and for some long time now, parting tools have tended to take the form of a tool-steel blade ground all over so that in section it is wedge-shaped. The blade is set in a forging of high-tensile steel which permits it to be extended as required by the machinist and locked in the holder at the point desired.

There have been many versions of the device, one of which is illustrated in Fig. 3.12, where the 'Eclipse' Parting Tool Holder is

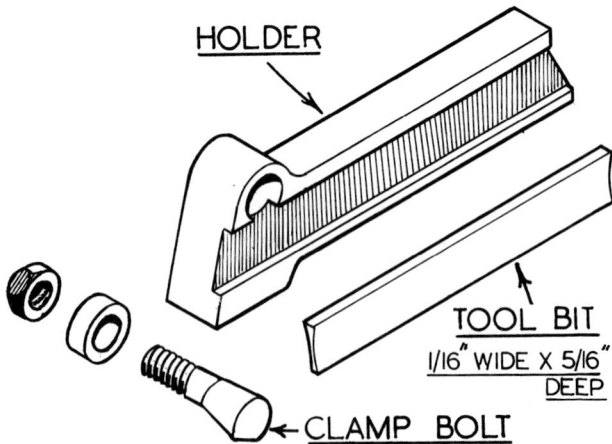

Fig. 3.13   Parts of the Parting Tool Holder.

depicted. Its component parts are seen in Fig. 3.13. The blade itself is of particular and important cross-section in that it is hollow-ground. The holder shown is the smallest in the 'Eclipse' range, designed for use in the smaller classes of lathe.

It should, perhaps, be explained that 'parting-off' is one of the more tricky lathe operations, particularly in light lathes where lack of rigidity may cause the tool to dig into the work. It is essential that cutting should be free with the chip leaving the work cleanly, and to aid this requirement the tool is given some top rake to make the chip curl well away from the kerf in which it is produced.

CHAPTER FOUR

METHODS OF HOLDING THE WORK

As we have seen, the only method of holding work in the very early lathes was to set it between centres and drive it in one of the ways described in Chapter 1. Obviously, as technology made progress, the limitations imposed by this method were felt increasingly, so it was not surprising that means were sought of holding the smaller and more complex parts that needed to be machined.

The Faceplate

Probably the earliest device for mounting work is the faceplate illustrated in Fig. 4.2. It is screwed directly to the nose of the mandrel and work is attached to it by clamps secured by bolts passing through slots

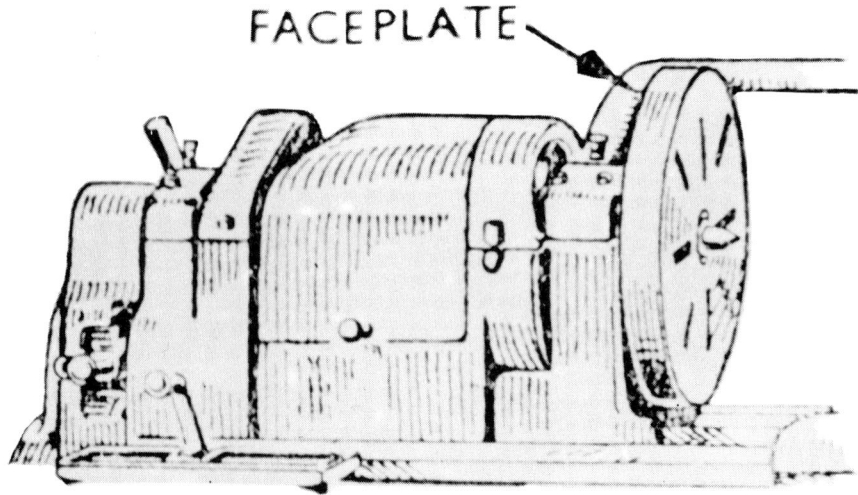

Fig. 4.1   The Faceplate.

Fig. 4.2   The Faceplate.

cast in the plate itself. The wood turner may well have been the first to use this method of holding work, and he still does when wooden dishes and the like have to be turned.

The faceplate can be furnished with dogs or jaws that enable it to be transformed into a simple chuck capable of holding irregular work, Fig. 4.3. The IXL lathe, available from Germany shortly after the 1914–18 war, had a well-developed composite faceplate of this nature.

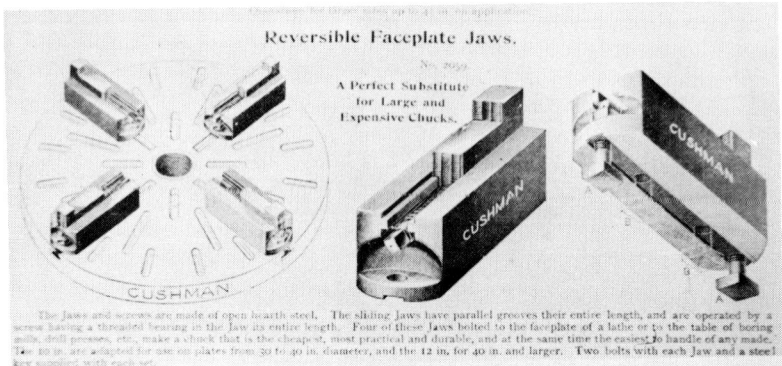

Fig. 4.3   Faceplate fitted with adjustable jaws.

The word 'chuck' has been used in connection with the device just mentioned. The term is held to mean a piece of equipment provided with a number of jaws that may be caused to grip the work, collectively or independently, and hold it while a turning operation is performed upon it. For the most part chucks are secured to the nose of the lathe, though some types are housed inside the mandrel itself.

Chucks—The Bell Chuck

The first stage in the development of the chuck is that illustrated in Fig. 4.4. It consists in providing a hollow body that can be screwed to the nose of the lathe, this body being called the 'bell' hence the name of the chuck. The bell has a number of set screws placed around its periphery so that any work placed within the device can be adjusted to run as truly as possible and then held firmly whilst turning operations are carried out. Bell chucks were evidently much in use at the early part of the 19th century for amongst other accessories used on the lathe, the 'Encyclopaedia of useful Arts' published in 1812, shows the device ready for mounting on the mandrel nose.

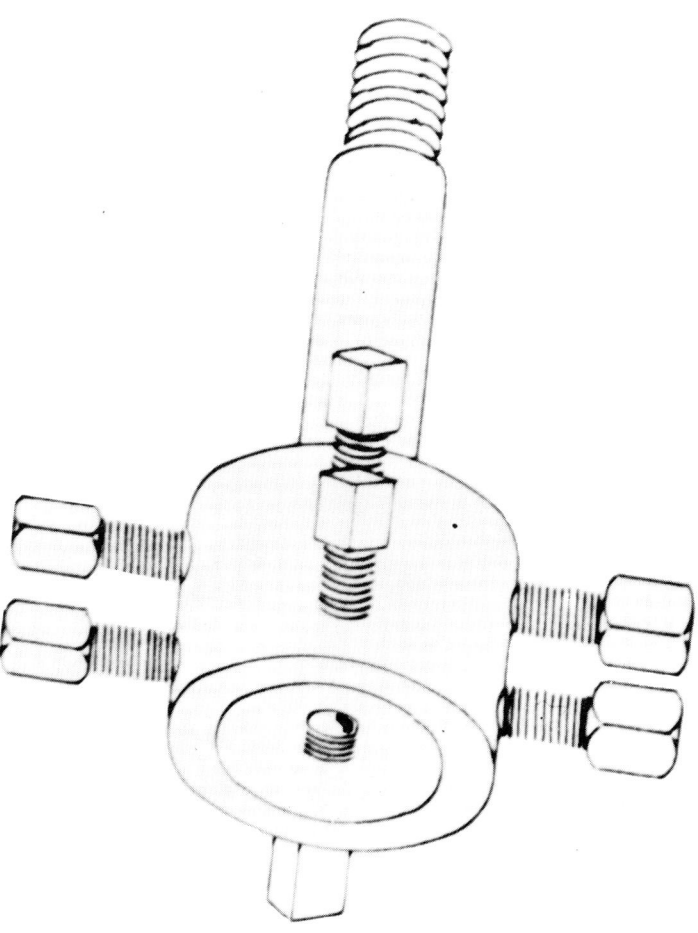

Fig. 4.4   The Bell Chuck.

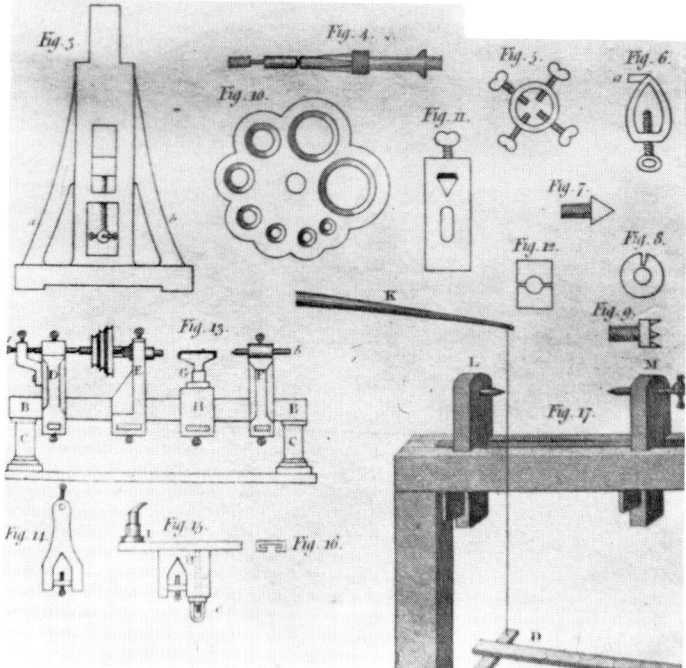

Fig. 4.5   The Bell-Chuck. See "Fig. 5."

The 4-Jaw Independent Chuck

The next logical stage in chuck development was the production from
America of the 4-jaw independent chuck. An example, taken from a
well-known tool merchant's catalogue of 1912, shows the Cushman
Company's chuck and the simple mechanism used in its construction.
As with other chucks of a like nature it is intended to be used with an
accurately machined backplate which screws to the lathe nose and is
fitted into the recess seen in the body of the device. Four bolts or
screws, passing axially through the body, secure it to the backplate. An
example, illustrated in Fig. 4.6, shows a modern 4-jaw chuck fitted to a
light lathe.

The Self-Centring Chuck

Setting work in the independent chuck is a somewhat slow-moving
procedure, so it is not surprising that some way needed to be found of
rapidly gripping the bar material then being used extensively in

Fig. 4.6   Modern 4-Jaw Chuck fitted to a light lathe.

industry. The Cushman Company of America, who had previously produced the independent chuck, introduced a self-centring chuck, having jaws that could be opened or closed by means of a T-handled key engaging pinions set in the body of the chuck. The pinions themselves mesh with a circular rack cut on the back of the scroll plate seen in Fig. 4.7.

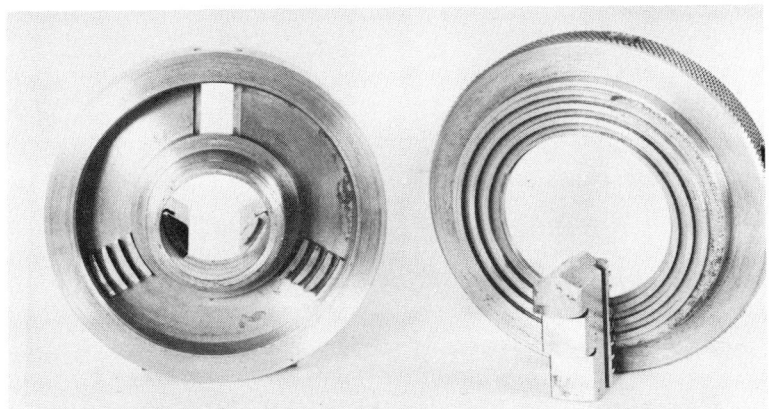

Fig. 4.7   The Scroll Plate and a Chuck Jaw from a lever operated Scroll Chuck.

Some of the earlier self-centring chucks, and in particular those intended for use in the amateur field, were lever operated. In point of fact the parts illustrated in Fig. 4.7 were photographed from a chuck of this type.

The lever, really a tommy bar, was inserted in holes formed in the periphery of the scroll plate. This allowed the operator to exert a reasonably firm grip in any work placed in the chuck, in particular parts made from wood. The author well remembers, as a young man, using such a chuck fitted to a very old treadle lathe.

Fig. 4.8   A Key-operated Self-Centring Chuck.

The scroll engages teeth machined on the back of each jaw so that, when the scroll is turned, the jaws advance or retire along the tenons formed in the chuck body. These may be seen in the illustration Fig. 4.8.

Mounting 4-jaw and Self-Centring Chucks

Mention has already been made of the backplate as a method of supporting chucks. It is vital that the backplate should fit both the lathe mandrel nose and the abutment face of the chuck itself. Accordingly the lathe manufacturers take great care that the machining of the backplate is as perfect as possible. Their first job is to set the backplate

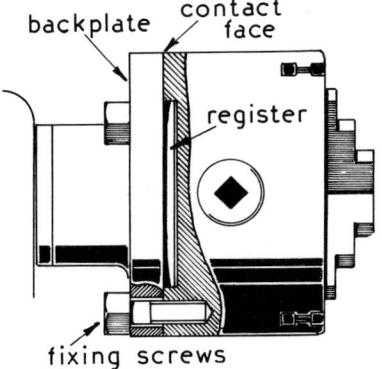

backplate   contact face

register

fixing screws

Fig. 4.9   Method of fitting the backplate to a chuck.

on the mandrel nose, making it a firm fit without shake, then with the backplate in place, a register is machined on the plate upon which the chuck fits firmly. The details of the procedure are illustrated in Figs. 4.9 and 4.10.

Where large lathes are concerned it is clearly not practicable to screw the chuck on to the mandrel nose, if only for the reason that the weight of the chuck itself would make this a difficult operation. Instead

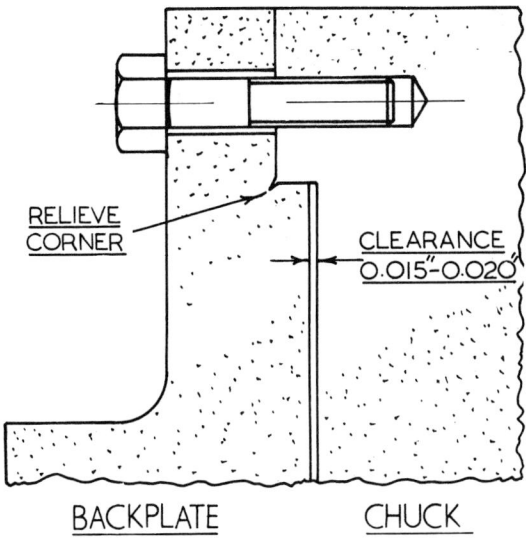

RELIEVE CORNER

CLEARANCE 0.015″–0.020″

BACKPLATE   CHUCK

Fig. 4.10   Method of fitting the backplate to a chuck.

the mandrel is furnished with an integrally machined flange to which chucks can be bolted directly after having been lifted into place by a crane.

An example of this arrangement as applied to the Dean Smith and Grace lathe is illustrated in Fig. 4.11.

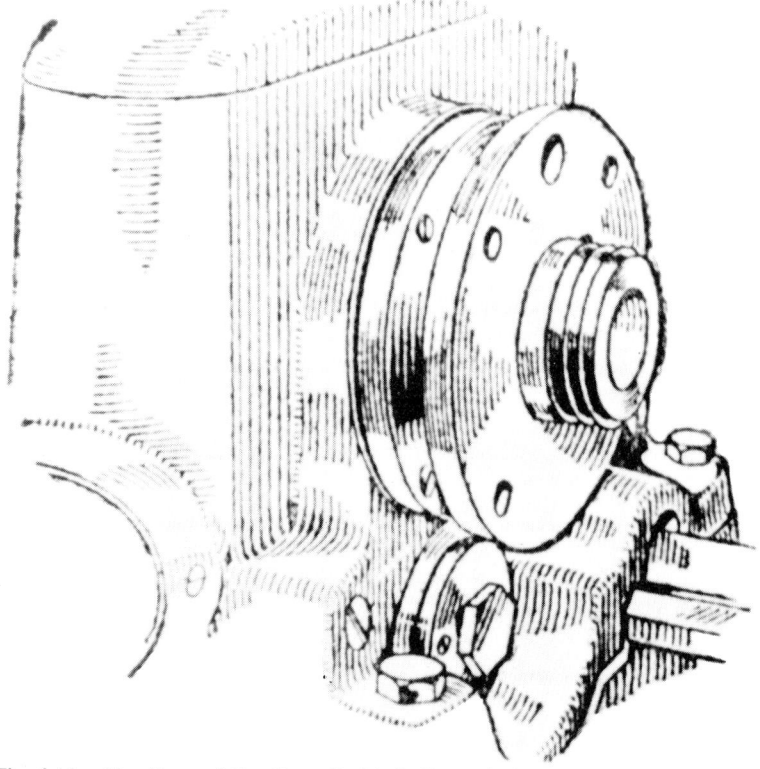

Fig. 4.11   The Nose of the Dean Smith & Grace Lathe.

The Combination Chuck

An attempt was made to combine the properties of the independent and self-centring chucks into one unit called, for obvious reasons, the combination chuck. Examples of the device appeared in tool merchants catalogues early in the 1900 era. But they were somewhat complicated mechanically so were probably relatively expensive to make. At all events they do not appear to have found much favour, industry at all events preferring to use separately the simpler and more robust chucks to which they had become accustomed.

Collet Chucks

The Collet chuck, illustrated in Fig. 4.12, is used principally in watchmakers lathes. Essentially this chuck is a tube split three or more ways for part of its length. It is furnished with an angular nose so that, when it is drawn into the hollow mandrel of a lathe having a corresponding internal cone, it will contract and grip work placed within it.

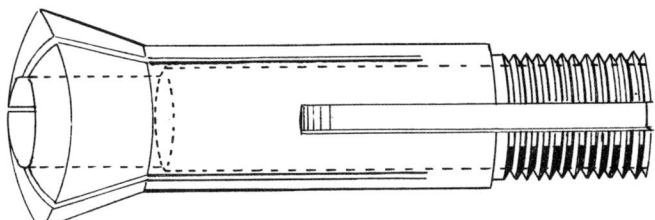

Fig. 4.12   The Collet Chuck.

Collet chucks are made to a high degree of accuracy. This enables work to be removed and replaced with the certainty that it will run true. In instrument lathes the chuck is closed by a draw-in spindle passing through the mandrel and turned by a hand wheel at the opposite end of the mandrel nose. The arrangement is depicted in he illustration, Fig. 4.13. It is, perhaps, worth noting that collet chucks need treating with great care if they are to remain accurate. Unfortunately, in industrial circles this does not always happen, for both oversize and undersize work is often put into collet chucks, and this they will not tolerate and still remain true. As one eminent authority, the late George Adams has said, 'There is really no latitude in a collet chuck. It will only take just that size of cylindrical work it is ground or lapped out to take.'

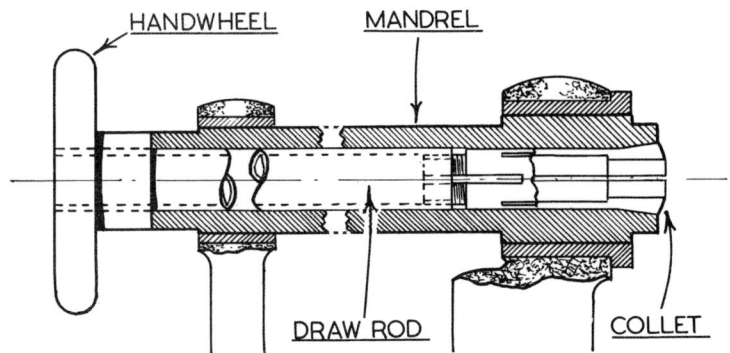

HANDWHEEL   MANDREL

DRAW ROD   COLLET

Fig. 4.13   The arrangement of Collet Chucks in a Lathe Mandrel.

Fig. 4.14   A set of Collet Equipment for a Centre Lathe.

Collet chucks fitted to production lathes are often closed by a cap, for the most part lever operated, the collet taking the form shown in Fig. 4.14 where a set of equipment for use with a simple centre lathe is depicted. Here the chuck is closed by the cap seen on the right of the illustration being forced into the adapter seen at the left as the cap is screwed home. The alternative set-up for lever operation is seen in Fig. 4.15.

Fig. 4.15   Lever operation for the Collet arrangements of a production Lathe.

The Myford Patent Colllet

An interesting development in the field of collet chucks is the patent collet made by the Myford Engineering Company of Beeston, Nottingham. This has been designed to fit the No. 2 Morse Taper seating

Fig. 4.16   Myford Collet Chucks.

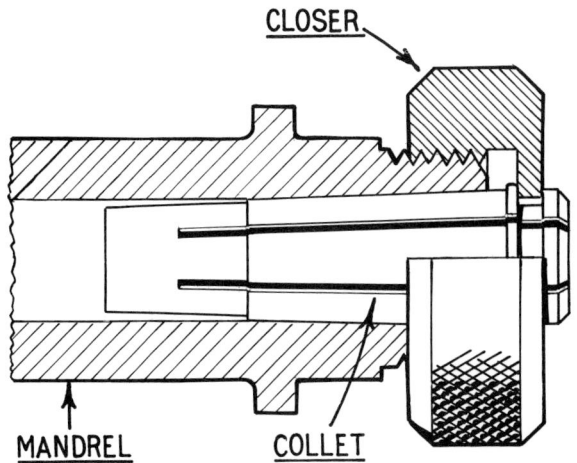

CLOSER

MANDREL   COLLET

Fig. 4.17   Arrangement of Myford Patent Collet in a Lathe Mandrel.

machined in the bore of the lathe mandrel itself. The collet is forced up the taper by a ring nut running on the thread of the mandrel nose. A series of Myford collet chucks together with the ring nut used to operate them is seen in the illustration, Fig. 4.16. The ring nut engages a groove machined in the nose of the collet so, when changing the ring nut from one collet to another, the collets need to be collapsed so that the ring can be slipped over them easily. The tool used for the purpose is that seen in Fig. 4.16. It consists of a tube into which the collet can be pushed to collapse it and a plunger to force the collet from the tube once the closing ring has been slipped in place.

CHAPTER FIVE

CAPSTAN AND AUTOMATIC LATHES

TOWARDS THE END of the 18th century the interchangeability of machined parts became an important factor. In 1785 a gunsmith named LeBlanc suggested that the production and repair of muskets would be much accelerated if their component parts could be made to a standard that enabled them to be interchanged readily.

By 1798 the requirement of the armed forces in many countries was leading to the placing of contracts for considerable quantities of small arms so interchangeability of their parts for repair purposes, to say nothing of their rapid production, was very important.

At this time an order for 10,000 muskets was placed with an American firm followed, in 1812, by an additional 15,000 weapons. These were all made on special purpose equipment, doubtless, including lathes capable of repetitive machining.

Lathes designed for rapid production are usually classed as capstan lathes, though a subdivision, the turret lathe, is often used for the larger components. The difference between the two lies in the location of the 'turret' that carries the tools used during the machining process.

In the capstan lathe the tool turret is mounted on a separate slide set on the bed of the lathe and capable of adjustment and positional selection. In this way each tool is brought into play at the time and in the sequence required. The Turret lathe, however, has no separate tool slide so carries its turret on the cross slide. It is in fact and for the most part, a centre lathe equipped with additional equipment to allow the production sequence to be followed.

From the manufacturing standpoint the great advantage of the capstan lathe is that, once it has been set by an experienced operative called, for obvious reasons, a 'setter', the machine can be handed over to an unskilled man who has little more to do than see that the tools in the capstan are applied in the correct order and are fed into the correct depth.

The first requirement is taken care of by the machine itself in the mechanism employed for turning the capstan and so is the second by means of stops working in step with each station of the capstan. The operative is thus left to ensure that, at most, he has read the index of the cross slide correctly. But even here a stop may be fitted to ensure that the operative feeds the cross slide into the correct depth without having to consult the index. The turret lathe, on the other hand, is a more complex machine that is unlikely to be manned by an unskilled workman but will usually be in the hands of a skilled 'setter operator' who, as his title indicates, is capable both of setting and operating the machine.

By 1840 several turret lathes were in use in America, these seem to have been of English origin, a patent for this type of lathe having been taken out in England sometime before this date.

By 1861 several major improvements to capstan and turret lathes had been introduced, in particular a device for feeding and gripping bar material whilst the lathe spindle was rotating and an automatically rotating turret with ratchet-and-pawl indexing mechanism, this last being developed in America.

Fig. 5.1   The Myford Lathe adapted as a Capstan machine.

A typical modern application of the capstan principle may be seen in Fig. 5.1 and Fig. 5.2 where the Myford Engineering Company's adaption of their centre lathe is depicted. In the first illustration the capstan attachment is seen assembled on the lathe bed together with a cut-off slide fitted with tool posts for front and rear working. In the second view the capstan attachment is depicted removed from its base

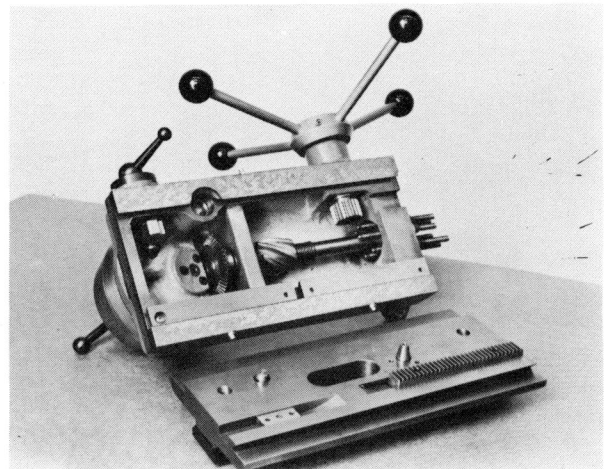

Fig. 5.2   The Capstan Attachment inverted.

plate and inverted. In this illustration the rack and pinion used to operate the capstan slide is illustrated together with the apparatus used to index the capstan head and set the stops appropriate to the several stations of that head.

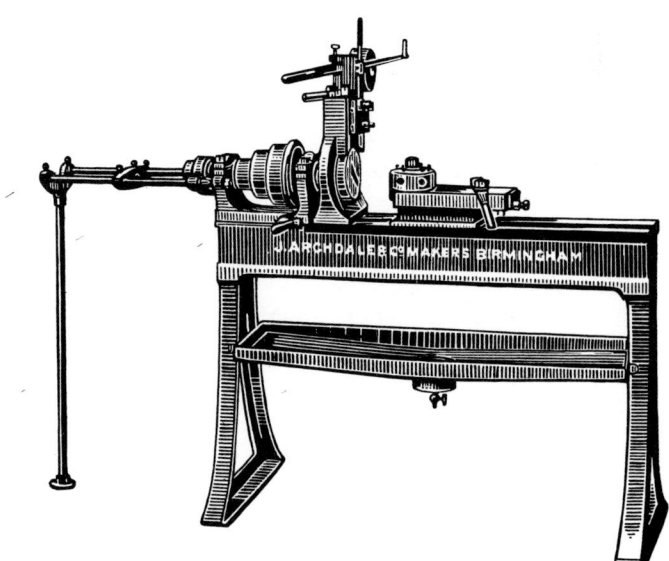

Fig. 5.3   Archdale Capstan Lathe 1884.

Both of these devices have been fundamental to the success of the capstan lathe; the first avoiding the loss of time involved in stopping and starting the mandrel whilst the second ensures that all tooling is presented to the work accurately and in the correct order.

A typical capstan lathe made by James Archdale about 1884 is illustrated in Fig. 5.3. This machine had all the essential features in such a lathe, having a bar feed and support system, a capstan slide and cut-off mechanism, this last set vertically on a casting above the chuck holding the bar material. The headstock was provided with a 3-step cone pulley and was intended for driving from the machine shop lineshafting through a counter shaft provided with a complimentary cone pulley.

Tooling

Apart from the equipment carried on the cross slide, the tooling used in both capstan and turret lathes differs materially from that employed on the centre lathe. The tools mounted in the capstan lathe are designed to be fed axially to the work and are made so that they can be adjusted to finish sizing the work at a single pass. This requirement applies equally to turning tools as well as to those intended for knurling or screw threading.

The Automatic Lathe

The requirements of mass production inevitably led to considering the possibility of making the lathe automatic.

In the first instance the class of work envisaged was the making of parts from bar material. The basic machine, therefore, was the capstan lathe which was suitably modified to provide means of operating the bar feed mechanism automatically as well as opening and closing the collet chuck gripping the material to be machined. Additionally all the tools required in the operation needed to be fed to the work in the correct sequence, to the right depth and for the distance set out in the working drawings.

The impetus for the introduction of the automatic lathe came from America where the supply of skilled labour was limited and any device to conserve it highly desirable. The early American automatic lathes seem to have followed the pattern introduced into England by C. M. Spencer in 1873.

These lathes were originally used to make screws, a commodity, in quantity, much needed industrially. As a result early automatic lathes were known as screw machines. But experience with them soon demonstrated that they could be employed for making a variety of other components.

The requirements of general machining has led to the development of automatic lathes of widely differing types. So much so that it would be impossible to cover all these in a single chapter. One can, therefore, only hope to give a broad outline of the general position.

As has been said, in automatic lathes, all the motions needed in a normal capstan lathe have to be performed automatically. From the outset these movements have been made by levers, actuated by cams mounted on a shaft or shafts set on the side of the machine. A typical arrangement is illustrated in Fig. 5.4.

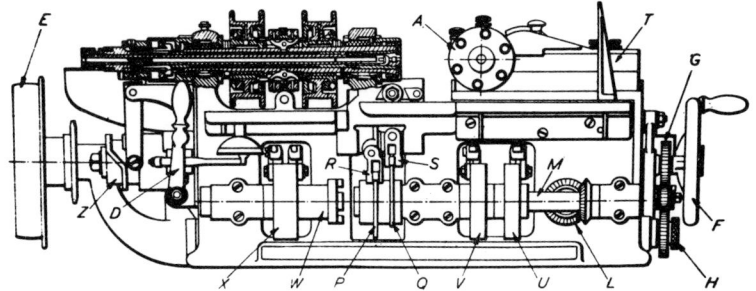

Fig. 5.4 The Brown & Sharpe Automatic Lathe.

It was also necessary to provide means of selecting headstock spindle speeds to suit the work in hand. In early Brown & Sharpe automatic lathes, two spindle speeds were available. These were controlled by clutches engaging the driven pulleys mounted on the spindle as may be seen in the illustration Fig. 5.5. High speeds were obtained 5,000 r.p.m. and 1,450 r.p.m. for example; these needed a copious flow of lubricant to the spindle bearings and this was supplied under pressure through the pipework seen in section below the headstock assembly.

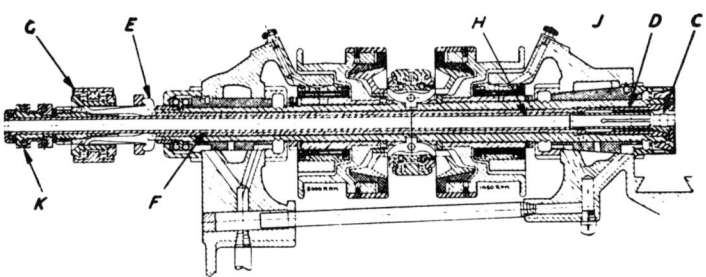

Fig. 5.5 The Brown & Sharpe Automatic Lathe Mandrel.

The Petermann Automatic Lathe

Mass production lathes so far described have been of the fixed head-stock type. Such machines, for fairly obvious reasons, are not suitable for the manufacture of long slender work such as the spindles used in electric light meters and parts of a similar nature. These need a machine having a headstock that can slide along the lathe bed carrying with it the bar material from which the parts are to be made. The bar is passed through a stationary bush, the same size as the bar itself, around which are grouped radially such tools as are needed for the actual turning operation. These tools are set in a series of cross-slides and are provided with mountings capable of micrometer adjustment so that work may be turned to close limits.

The rotating bar material is pushed through the bush, which supports it, and past the turning tools; these as has been said, can be adjusted to give the work the varying diameters required. In this way parts of considerable length, and of somewhat slender form, can be machined speedily and with great accuracy.

When it is necessary to drill the work or to form a thread upon it this is carried out by tools mounted on a separate slide or slides co-axial with the main headstock with spindles capable of being driven under power. The tools are brought to bear on the work as and when required, their speed of rotation being controlled automatically.

The method of use for threading work in the Petermann lathe is somewhat interesting because it is carried out with both the work and the threading die rotating. It will be appreciated that in cutting a thread the die itself only rotates comparatively slowly. In order to achieve these relatively slow speeds both the work spindle and the die are set to rotate in the same direction with the die revolving slightly the faster of the two. In this way the die travels slowly along the work cutting the thread as it goes. When the thread has been formed to the correct length the speeds of the two spindles are interchanged causing the work to withdraw from the die.

In all automatic lathes provision has to be made for removing the finished component from its parent bar material. The tool for this purpose is almost always carried in a separate post attached to an independent slide operated by a cam forming part of the lathe's general tool-timing mechanism. A typical arrangement is that of the Brown & Sharp auto seen in Fig. 5.6, where the actuating mechanism for the front and rear cross-slides of that machine are depicted.

Early automatic lathes had but a single work spindle. A little consideration, however, demonstrated that more effective use would be made of the tools used in machining, as well as materially increasing the rate of production, if the number of work spindles could be

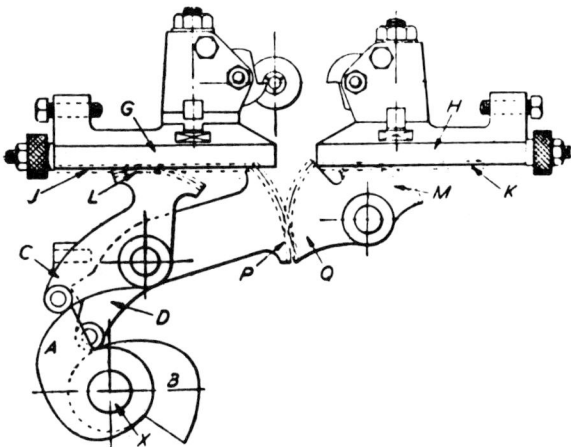

Fig. 5.6   Brown & Sharpe Cut-off Slides.

increased. Accordingly a design was worked out by the New Britain-Prentice Company of America in which four spindles were geared together carrying bar material that could be worked on by tools mounted in a turret set co-axially with the work spindles. The turret had provision for indexing to enable the tools to approach the work in the correct order. The machine itself is illustrated in Fig. 5.7 where the main assemblies can be seen in some detail.

As has already been said the whole subject is complex and can only be covered very sketchily in a single chapter. However, there may still be available in the public libraries, for they have long since gone out of print, two excellent short books by the late Edgar Westbury that cover the subject admirably. The titles are: 'Capstan and Turret Lathes' and 'Automatic Lathes and Screw Machines' respectively.

Both subjects are treated from a practical standpoint that may very well appeal to present company. There is no lack of illustration so readers may expect to obtain all the information they may need.

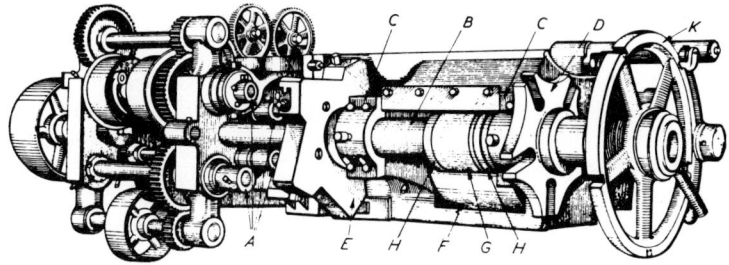

Fig. 5.7   4-Spindle New Britain Automatic Lathe.

CHAPTER SIX

THE MILLING MACHINE

THERE IS LITTLE DOUBT that the milling machine owes much to the lathe for its development. Clockmakers had, for some years, used their lathes to mill such work as needed this treatment, a practice that still survives in the amateur field as we shall see later.

In this connection simple equipment for the cutting of the teeth in gear wheels used with clocks and scientific instruments was available by 1789, the first reference to a milling machine being in a French encyclopaedia about 1772.

If the advantages to be gained from the practice are to be fully realised, the technique of milling, for the most part, requires a machine that can provide a wide range of spindle speeds.

Richard Roberts (1784–1864) had designed and built in 1817 a lathe fitted with back gear, so the arrangements needed to provide a reasonable range of spindle speeds were already in existence. The illustration of the early milling machine depicted in Fig. 6.1 shows that it was furnished with back gear together with self-acting mechanism enabling the work table supporting a vice to be traversed under a cutter mounted on the machine spindle. Depth of cut was adjusted by the simple expedient of raising or lowering the spindle itself and, of course, the re-setting the meshing of the back gear which this involved and for which provisions were made. Cutters of various types could be used with such a machine, a subject with which we will be dealing later.

By the year 1821 Roberts was advertising himself as 'a maker of screws, division plates, bevel, spur and worm gears up to 30 inches diameter'. Field, who was Maudslay's partner, had visited Roberts' workshop and noted that 'these gears were cut by milling' and that 'the cutters were made from steel plate cut to the shape of the tooth'. Evidently specialisation had arrived early. The range of products that Roberts was advertising in itself presupposes a considerable technical advance. It is somewhat disappointing, then, that Field seems to have confined his remarks on cutters to those quoted, and has left us guessing about the remaining processes in their production

64

Fig. 6.1   An early milling machine, the design shows many traces of evolution from the lathe.

Roberts himself was originally an employee of Wilkinson, the inventor of the boring machine. He was an outstanding designer who, so far as England is concerned, was responsible for most, if not all, of the pioneering work on the milling machine. He seems also to have taken no small part in the development of other machine tools.

Nasmyth (1808–1890) now comes into the list of those concerned in the development of the milling machine. He designed and built a self-acting machine for milling the faces of hexagonal nuts. From this it will be seen that the evolution of special purpose machines was already in progress, a development that, in one direction or another, has been going on ever since.

By 1850 the practice of milling had become firmly established as a manufacturing process, and many firms were making the necessary equipment. For workshop management had begun to see that, when producing selected components, the milling machine had many advantages over other machines such as the shaper and planer which could be freed for other and more suitable work.

In America, meanwhile, development was taking place independently, the work being based, according to some authorities, on results originating from England; and it is on record that a very crude milling machine had been installed in a gun factory at Middletown in Connecticut during the year 1818.

This machine designed by Eli Whitney, is depicted in outline in Fig. 6.2. At the time that Whitney produced his design the hand file was the principal tool used by gunsmiths, and the new machine was employed only as a roughing tool to lessen the drudgery of the handwork needed to produce gun parts, leaving only filework to finish them. The machine was applied to plane surfaces only, no curved work or profiling being possible with it.

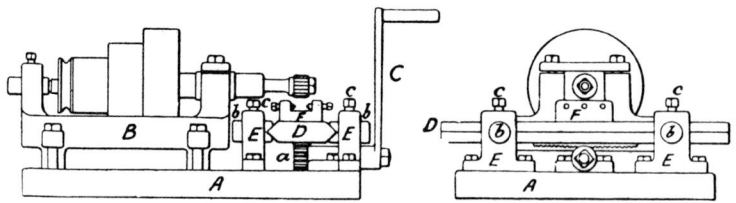

Fig. 6.2   Eli Whitneys milling machine.

Whitney's machine consisted of the following parts: a bedplate 'A', a casting measuring 24in. × 18in. × 2in. thick. The headstock 'B', taken from an old centre lathe and bolted to the bed, carried a spindle with a square-tapered socket to receive the arbor upon which the cutter itself could be mounted. The cutter was plain with teeth formed by hand filing and was about $1\frac{1}{2}$in. in diameter with a 1in. face. The cone pulley attached to the spindle was made of wood with steps some $2\frac{1}{2}$in. wide, the largest being about 8in. diameter.

Work was traversed under the cutter by the hand crank 'C' turning a pinion engaging the rack fixed to the underside of the work table 'D'. The table edges were V-shaped to fit notches cut in the plugs 'b', 'b' carried in the uprights 'E', 'E' bolted to the bed of the machine. When wear took place the plugs could be adjusted and locked by the set screws 'c', 'c'.

The machine remained for some years in the form illustrated but later an automatic work table feed was added. Work was held in the vice 'F' but no means of adjusting the depth of cut was provided other than by introducing packing under the work itself.

Obviously such a practice was exceedingly wasteful of time, so it is not surprising that the next important development lay in the direction of adjusting the depth of cut in a simple but positive manner. Initially this was effected by replacing the fixed spindle bearings by an arrangement that allowed them to be adjusted vertically, thus doing away with the troublesome procedure of having to insert packing under the work each time a new cut was taken. The machine already illustrated in Fig. 6.1 has this arrangement.

Later a tailstock was added, this was also adjustable for height and added greatly to the rigidity of the machine as a whole.

Fig. 6.3 An early Lincoln Milling Machine.

D

Milling machines built in this form are known as Lincoln millers. Because of their simplicity, and rigidity, they achieved wide popularity. Indeed, for certain classes of work, machines constructed on these lines still retain it. An early Lincoln milling machine is depicted in Fig. 6.3 whilst a later form is illustrated in Fig. 6.4 showing how well the simple but rigid design has been retained. The Lincoln was introduced in 1854 and was the first machine to have had a screw movement to the table feed instead of the rack and pinion hitherto in use.

In 1861 Joseph Brown of Brown and Sharpe in America, invented the prototype of the universal milling machine, developing it from equipment used to machine the flutes in twist drills. Brown's first universal milling machine is illustrated in Fig. 6.5. The machine had

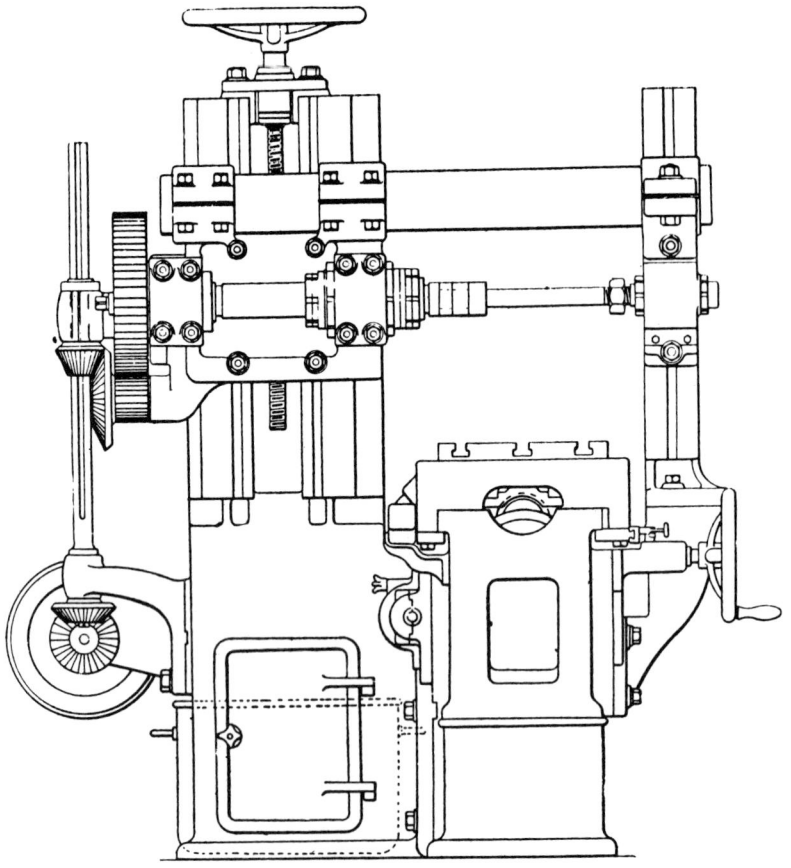

Fig. 6.4   A later type of Lincoln miller.

Fig. 6.5　Joseph Brown's first universal milling machine 1861.

no back gear and the spindle bearing assembly was fixed. On the other hand it possessed what all milling machines of a like nature now have, that is a rising and falling 'knee' supporting the work table and its slides enabling very accurate adjustment of cutting depth to be made. It also possessed a dividing head, the purpose of which we shall be discussing later.

In order to carry out work akin to the fluting of twist drills it is necessary to set the work table at an angle to the cutter, and it is this attribute that distinguishes the universal milling machine from the plain miller we have considered previously.

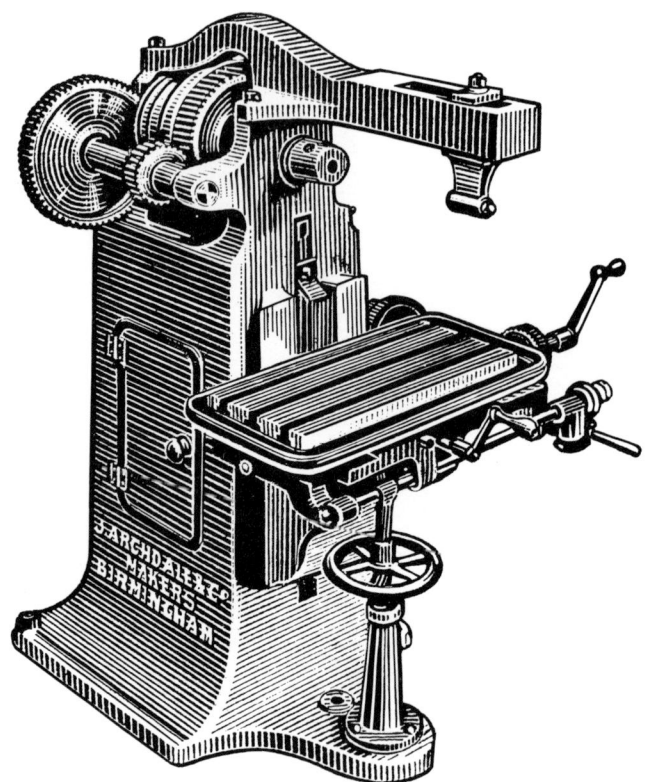

Fig. 6.6　A Double geared Horizontal Milling machine 1884.

The Vertical Milling Machine

During the latter half of the 19th century, the vertical milling machine, developed to some extent from the drilling machine, had been introduced to carry out, initially, such work as the machining of keyways in shafts. An early machine suitable for this purpose is illustrated in Fig. 6.7. Later it was realised that, provided enough rigidity was built into it, the vertical miller could be used for a very much wider range of work. The machine illustrated in Fig. 6.8 is of a more robust type, having provision for the reversal of spindle direction if required as well as self-acting mechanism for both the work table feed and the rotary table seen in place on the work table.

Up to the year 1904 milling machines were driven from the shop lineshafting using countershafts similar to those employed in connection with lathes. In 1904, however, Brown and Sharpe, who had already made many improvements to it, brought out a miller with single pulley drive and independently variable changes of feed and speed available through gearing built into the machine itself. In this way the milling machine was following the pattern set by the centre lathe and the way was opened for the completely self-contained machine with its own built-in driving motor and, later of course, independent motors and gearing for the table feed.

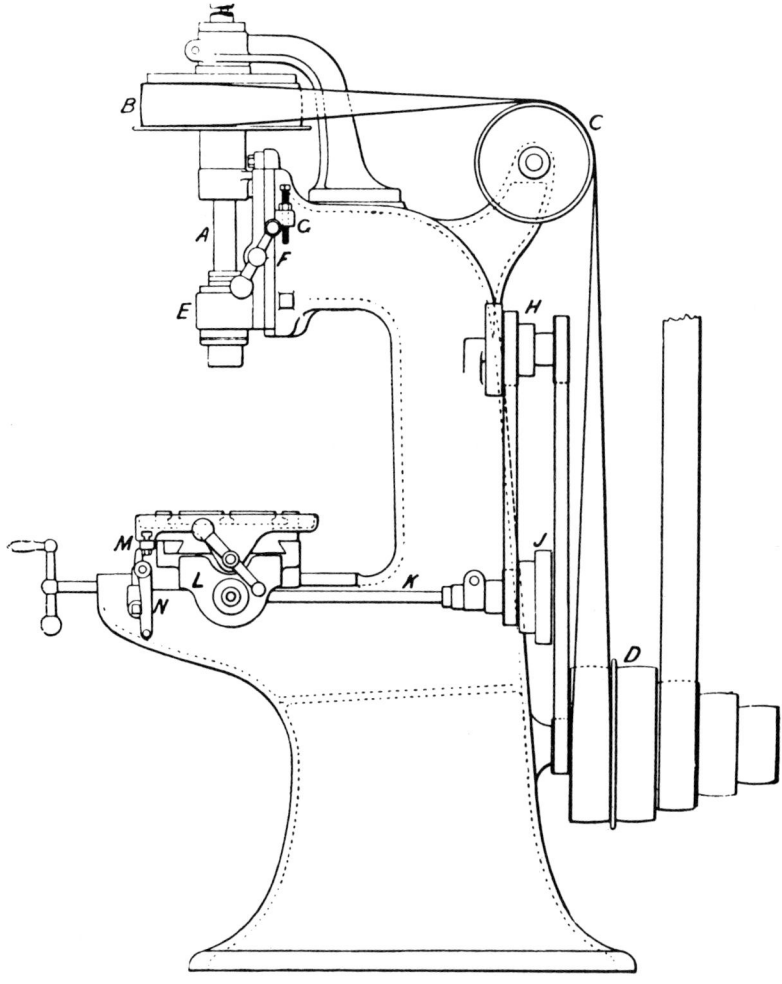

Fig. 6.7   An early Vertical Milling machine.

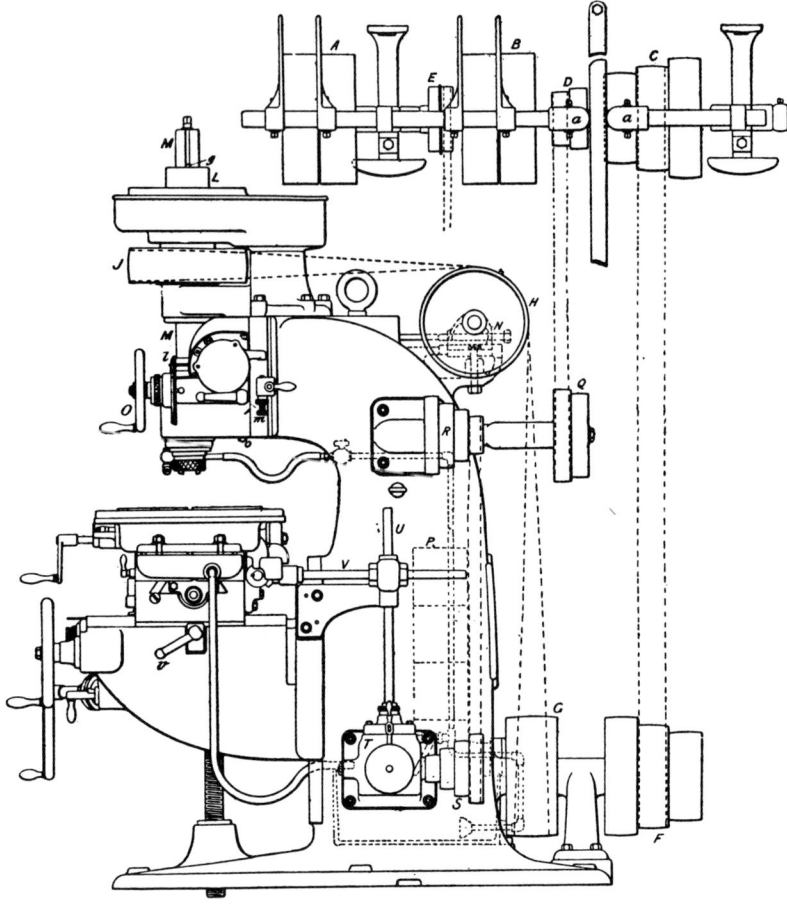

Fig. 6.8   A later and more rigid vertical miller.

The versatility of the milling machines mechanism, and the comparative ease with which work could be set up on it, soon led to attachments made for the machine enabling it to perform work which would otherwise need a transfer to another machine tool.

The two attachments depicted in Fig. 6.9 and Fig. 6.10, respectively, are examples in point. The first is a vertical milling attachment designed to be bolted to the face of the machine column and driven from the main spindle. The particular attachment illustrated follows common practice in these matters and is provided with a steady

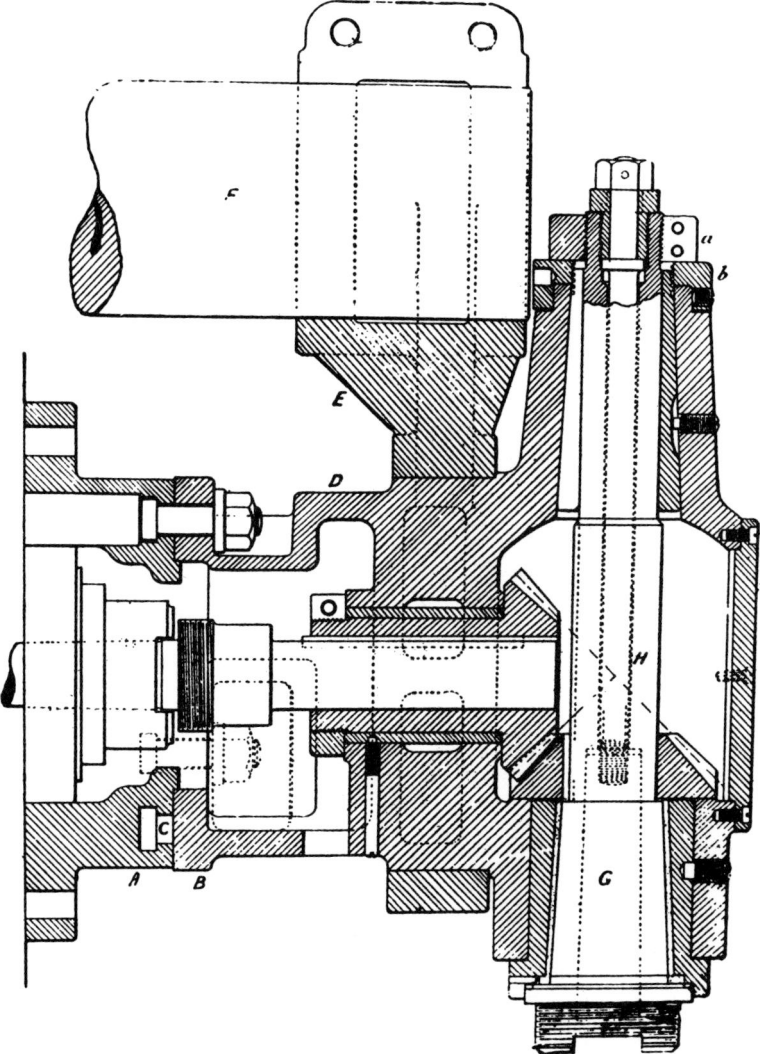

Fig. 6·9   A vertical milling attachment for the horizontal miller.

bracket 'E' that can be clamped to the overarm of the milling machine and is secured to the attachment's gearbox 'D'.

The second device is a slotting attachment mainly used for cutting internal keyways. It is bolted directly to the side of the spindle bearing housing and is driven from a multi-start worm 'B' fitted to the machine spindle 'A'. The worm engages the wormwheel 'C' on which a crankpin is mounted and this, in turn, operates the connecting rod 'D' transmitting movement to the slide assembly 'E' carrying the toolpost 'F'. There

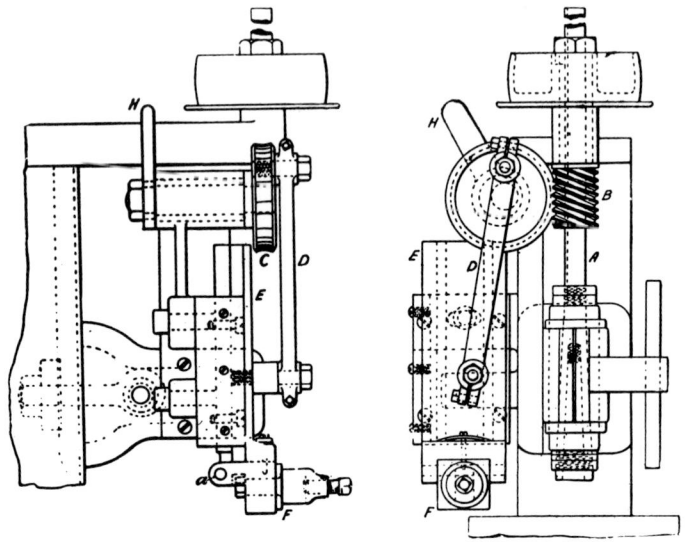

Fig. 6.10   A slotting attachment for the vertical miller.

is no means of adjusting the stroke of the tool, and the device as a whole requires a rising-and-falling work table in order to set accurately the position of the tool in relation to the work itself; although it is possible that, by either lifting or lowering the spindle bearing housing and moving the attachment as a whole, tool position could be adjusted.

The Dividing Head

Mention has already been made of the dividing head and the rotary table, equipment without which the versatility of the milling machine would be greatly reduced, so no account of the development of the miller can be concluded without some reference to these attachments.

Methods of dividing work into equal parts for machining were probably invented by clockmakers of the 18th century who needed this capability when cutting teeth in clock wheels. They made use of a simple division plate and index, the first being fixed to the face of the lathe pulley whilst the index itself was attached to a bracket on the headstock casting as seen in the illustration Fig. 6.11. The plates were provided with rings of equally spaced holes, the numbers usually selected being 96, 112 and 360, numbers that afforded the widest variety of divisions. In use, with a simple arrangement of this nature, the index pin is placed successively in those holes that provide the divisions needed. For example, when say 12 divisions are required the 96-hole ring in the division plate is selected and the pin placed in every eight hole.

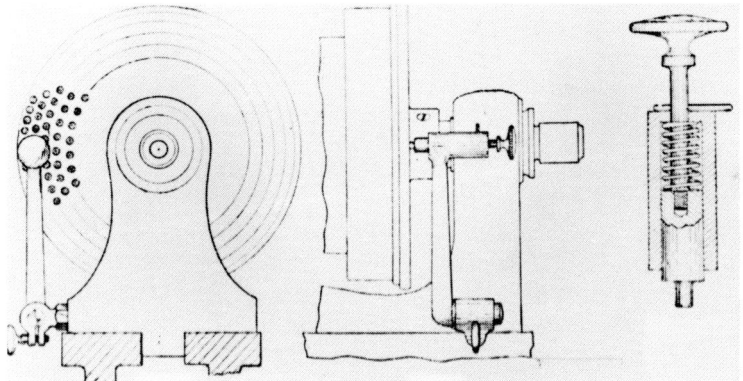

Fig. 6.11   Division plate and index fitted to a lathe headstock.

Clearly, though this method of dividing was accurate, it was a somewhat laborious proceeding that could not be countenanced when speed in production became a major consideration. This led to the introduction of the dividing head itself, though by whom is not clear. Richard Roberts may well have had a hand in it. As we have seen by the year 1821 he was advertising division plates, worms and worm wheels, all parts upon which the dividing head itself is based.

The device consists essentially of a worm and worm wheel geared together. The wormwheel itself is mounted on an arbor adapted to hold work that is to be machined, whilst the worm is attached to, or is machined on, a shaft fitted with an index pin engaging a division plate. The ratio of the worm and wormwheel is usually 40 : 1, that is to say, the worm makes 40 turns for each single turn of the wormwheel. Some dividing heads, however, work on a 60 : 1 ratio.

As an example of simple dividing using the 40 : 1 gear ratio we can suppose that a piece of work needs to be divided in 20 divisions. To do so, the worm shaft needs to make two turns for each division. Similarly, for ten divisions the worm shaft makes four turns, for eight divisions five turns, and so on.

From this it will be seen that when the number of teeth in the wormwheel are divided by the number of divisions required, and the result is a whole number, then the index shaft itself makes a whole number of turns for each division.

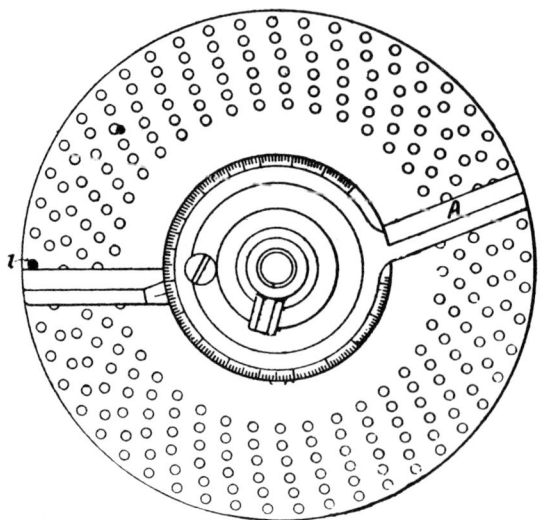

Fig. 6.12   Division plate and sector arms.

When on the contrary, divisions have to be made that are not equally divisable into the number of teeth in the wormwheel, the index shaft has to make a series of partial turns if the number of divisions is more than 40, or whole turns plus parts of a whole turn when the number of divisions required is less than 40.

As an example of the latter let us suppose that a gear wheel having 35 teeth has to be cut. Then the index shaft must make $\frac{40}{35}$ or $1\frac{1}{7}$ turns for each tooth division. A device has therefore to be added to the dividing head to ensure that the index shaft makes fractional parts of a turn accurately.

To ensure this the division plate, already referred to, is employed, this time mounted on the bearing housing of the index shaft. Division or index plates used on the dividing head have several rows of holes and there may be several interchangeable plates available for each dividing head. They are provided with sector arms, seen in the illustration Fig. 6.12, that can be set to embrace any desired number of holes, and then locked together. In addition the index pin itself can be adjusted to engage any ring of holes required.

To return to our example one seventh-of-a-turn can be indexed by setting the sector arms the correct number of holes apart using any circle of holes divisible by seven. If a 49-hole circle is selected then the arms would be set seven spaces or eight holes apart, with one arm against the index pin.

Fig. 6.13 A dividing head mounted on the table of a Brown & Sharpe Milling machine.

Then the index shaft is rotated one whole turn plus one seventh of a turn by carrying the index pin forward until it comes into contact with the second sector arm. The sector arms are then turned until the first arm is again in contact with the index pin when the whole process is repeated until all 35 divisions are complete.

In Fig. 6.13, a complete dividing head is seen mounted on the work table of a Brown and Sharp horizontal milling machine; this machine also carries an example of the vertical attachment we have been considering earlier.

Before leaving the Dividing Head one further type must be mentioned. This is the device used for differential dividing. In equipment of this nature not only does the index shaft rotate but so also does the division plate itself. The reason for this arrangement will be well known to the expert, other readers, however, would find little interest in a detailed description of differential dividing.

The Rotary Table

On the vertical milling machine the dividing engine is the rotary table. This device consists of a plate upon which work may be mounted; this plate is rotatable on a base fixed to the table of the milling machine. The plate, which is T-slotted so that work can be secured to it, has figures from 0 degrees to 360 degrees engraved on its edge, while the base is provided with an index line to allow the table to be employed for angular dividing. A typical rotary table is illustrated in Fig. 6.14.

Fig. 6.14   A typical rotary table,

The instrument makers vertical milling machine depicted in Fig. 6.15 demonstrates a typical application of the rotary table to a light machine. As will be seen the table is secured directly to the cross-slide, taking the place of the machine vice seen to the right of the illustration, though this, too can be affixed to the rotary table itself if need be.

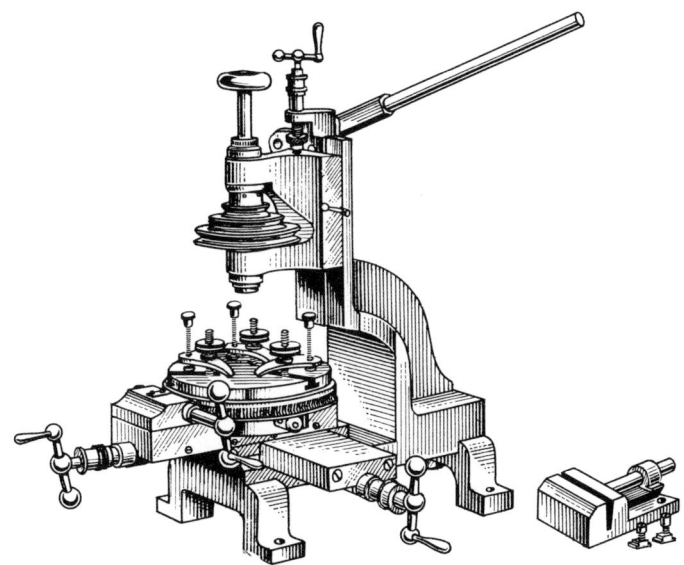

Fig. 6.15   An instrument makers vertical milling machine.

Milling Cutters

The cutters used in milling take two forms; the first of these, used in horizontal machines for the most part, is akin to a circular saw having a number of teeth set round a disc or hub that may be secured to the milling machine arbor. The second form is used in the vertical miller and has the drill for its parent; indeed one type of cutter in use today is known as a slot drill.

Until the year 1864 milling operations were performed either by 'fly-cutting' or by means of elementary multi-tooth cutters that could be sharpened in much the same way as a lathe tool. The fly cutter, which is illustrated in Fig. 6.16, comprises an arbor to fit the spindle of the milling machine, the arbor itself having provision for mounting single-point tools similar to those illustrated.

As the amateur worker, and even the professional, knows well the flycutter has the great advantage that quite complex forms can be implanted upon it by simple hand methods; on the other hand, having but a single cutting edge, the flycutter is comparatively slow in operation, a disadvantage that led to the introduction of multi-tooth cutters. But early cutters of this type also suffered a great drawback in that, before they could be sharpened, they had to have their temper drawn, and that they had to be hardened again before being brought into use.

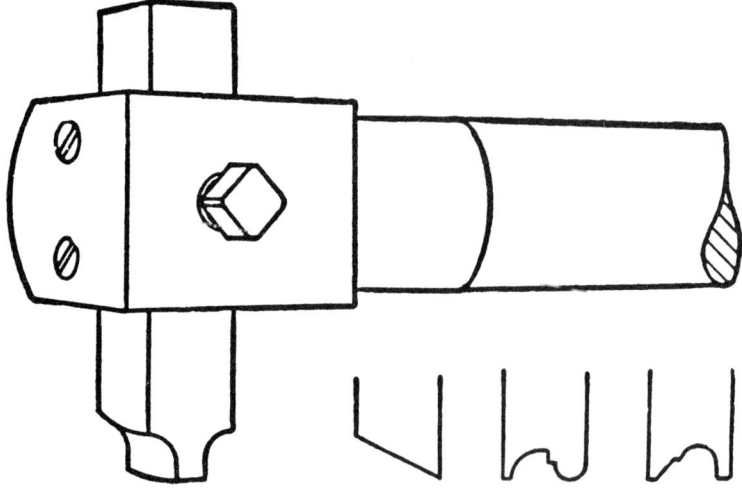

Fig. 6.16   Early fly cutter with arbor and some typical tools.

Development of the milling cutter in its modern form had, therefore, to await the invention of the emery wheel and its attendant equipment before the fullest use could be made of these cutters.

However, the first practical circular milling cutter appears to have been developed as the result of the clockmakers requirements in France, Jacques de Vaucanson 1709–1782 produced the cutter illustrated in Fig. 6.17. Historically this is said to be the first milling cutter ever produced and was probably used in some forms of attachment to the centre lathe. It did not possess teeth of the form now associated with milling cutters, but was more like a circular file with teeth resembling those of the 'Dreadnought' or 'Millenicut' file employed in connection with aluminium alloys.

Perhaps the most interesting point about the Vaucanson cutter is that it was designed to fit an octagonal shaft, the bore of the device being shaped correspondingly. In this way a positive drive to the cutter was assured. Today, except for small cutters and saws where friction

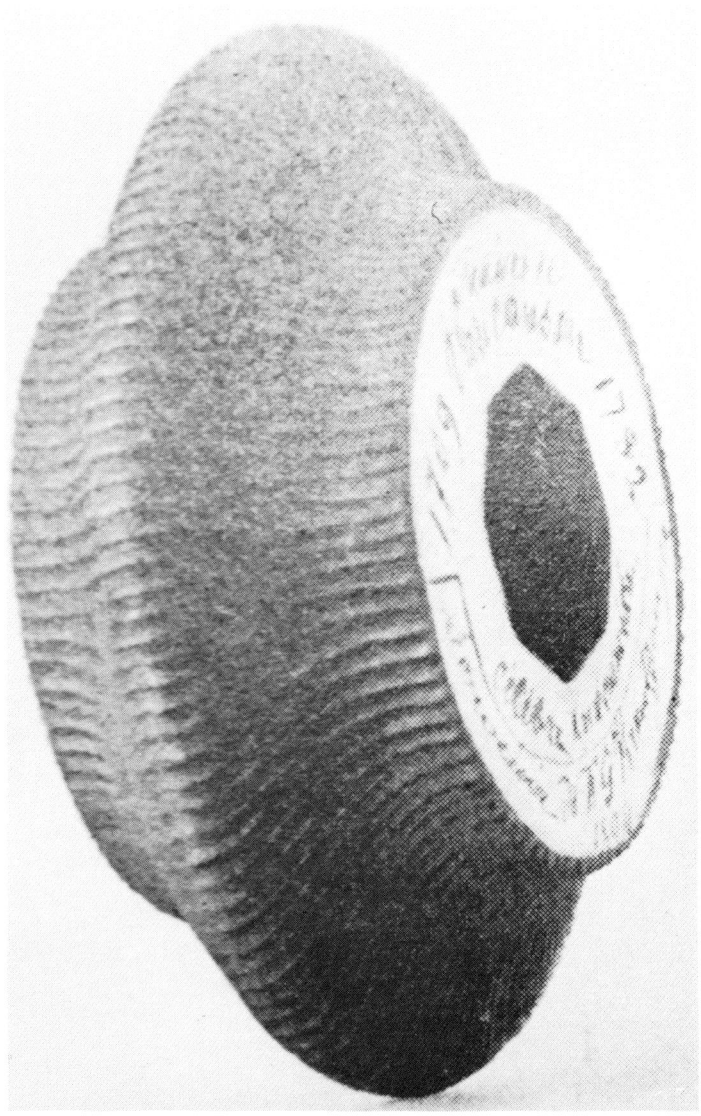

Fig. 6.17   The Vaucanson Cutter.

Fig. 6.18   J. R. Brown's form cutter.

alone is sufficient restraint against slip, milling cutters are provided with a keyway to engage a corresponding key forming an essential component in the cutter arbor itself. It seems unlikely that Vancanson's cutter can have run truly, for the teeth were formed by hand to an approximate gear tooth shape and so can have had but little accuracy. It was this demand for accuracy, particularly in the cutting of gears, that led J. R. Brown, of whom we have already heard, to the invention in 1864 of the form cutter depicted in the illustration Fig. 6.18. This cutter is one designed to produce gear teeth of involute form and has itself teeth machined by a relieving operation carried out, for the most part, in the centre lathe. As a result, and provided that the face of each tooth was ground so that its surface if projected would pass through the centre line of the cutter that is to say it was radial to it, then the form would remain constant however many times the cutter was re-sharpened. This was a notable advance, for not only had cutter life been materially extended, but re-sharpening itself had thus been much simplified.

Flycutting

Reference to "Flycutting" has already been made earlier where the practice, when related to the milling machine, has been described. Flycutting is a method of machining that has its origins in the lathe. Originally developed by the clockmakers for cutting teeth in gear-wheels it has been, and still is, used to machine surfaces on work mounted on the lathe cross slide. The amateur worker is mostly concerned with this particular practice because the lathe is a 'do-all' for him whereas the professional operative obviously has other and more versatile machines at his disposal.

Flycutting from the Headstock

When work is mounted on the lathe cross slide the flycutting tool is set either on the faceplate or in a special adapter caught in the chuck (see Fig. 6.19). Alternatively, it may be set in a boring bar mounted between centres.

The first of the two methods of mounting the cutter is used for the larger surfacing operations, whilst with the tool set in a boring bar or in an appropriate fixture already mentioned it is possible to machine concave surfaces.

An example of the first method is depicted in Fig. 6.20 whilst Fig. 6.21 illustrates the manner in which the third method can be applied.

Flycutting can also be employed to machine keyways in shafts gripped in a vice mounted on the lathe cross slide. This practice, of interest mainly to the amateur worker, is depicted in Fig. 6.22.

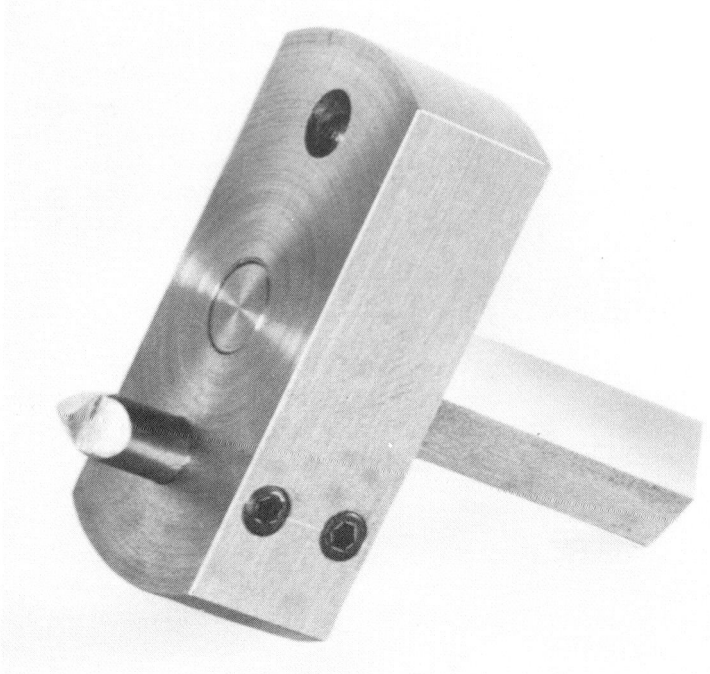

Fig. 6.19   Flycutting adapter for chuck mounting.

Flycutting with Saddle-Mounted Attachments

One of the oldest methods of carrying out milling operations is a frame set on the lathe saddle and driven for the most part, from an 'overhead' shaft set above the lathe and sometimes driven independently of it.

Milling in the Lathe

The subject of milling as a machining process cannot be dismissed without reference to the lathe and its use for the purpose. The first people to employ the lathe in this way were the clockmakers. They needed a ready means for cutting the teeth in their clock wheels, so they devised a cutter frame that could be mounted in the lathe toolpost and driven from overhead shafting. The work was mounted on the head-stock being supported when needed by the tailstock. A typical cutter frame of the period is depicted diagrammatically in Fig. 6.23 which

Fig. 6.20   Carrying out a surfacing operation with the flycutter.

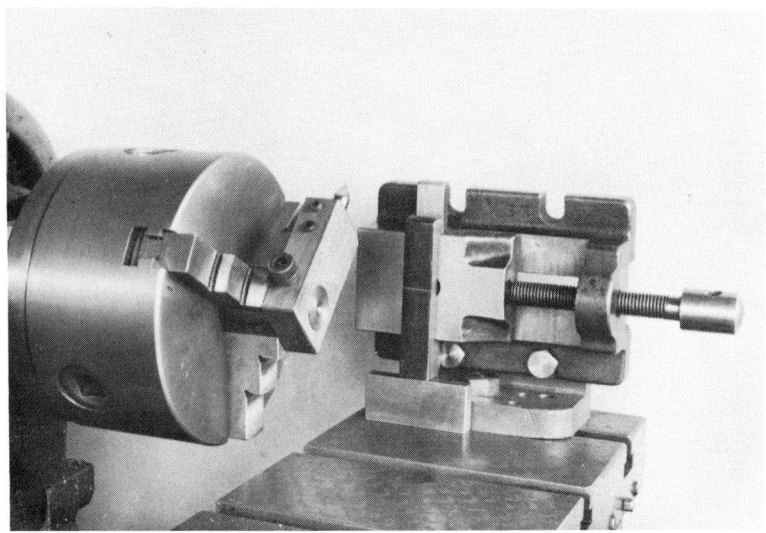

Fig. 6.21   Set up for machining a curved surface with a flycutter.

Fig. 6.22   Flycutting a long keyway.

Fig. 6.23   Diagrammatic representation of the cutter frame.

shows the method commonly used to support the cutter spindle. As will be seen the spindle, which carries a small pulley and a seating for the cutter itself, is mounted on a pair of hardened screws; one being coned to fit the corresponding female seating machined in the end of the spindle whilst the other is coned to accept the pointed end of the spindle.

A practical application of the system is illustrated in Fig. 6.24. This device was made by a working restorer of old clock mechanisms in order to produce the pinions and gearwheels needed in his work. The

Fig. 6.24   Practical application of the cutter frame system.

frame itself is built up from mild steel sections while the coned screws upon which the spindle runs are hardened and are provided with lock nuts. In this way the screws can be secured after adjustment to the running of the spindle. The frame is attached to a split mount that is itself set on a spigot forming part of the lathe top-slide.

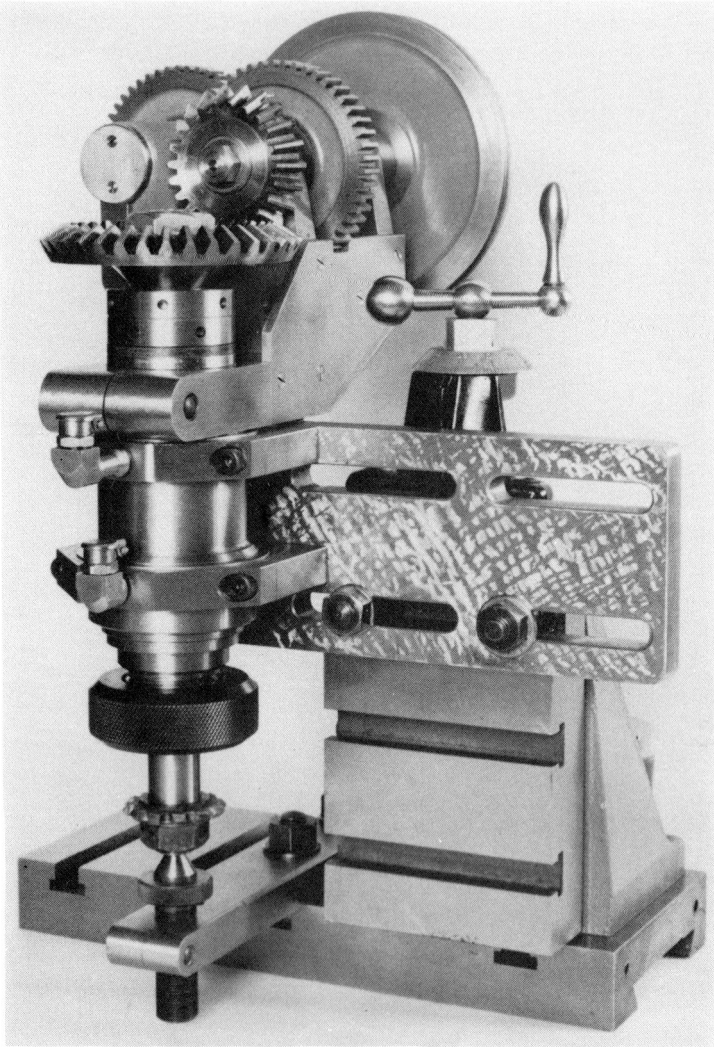

Fig. 6.25   The author's lathe milling attachment.

Fig. 6.26 The overarm.

The equipment illustrated in Fig. 6.25 is the logical development of the devices just described. It was made by the author in order to cut spur gears of varying pitches using form cutters of a type normally employed for the purpose industrially.

The spindle and bearings used are similar to and the same size as those of the $3\frac{1}{2}$in. centre lathe with which the milling attachment is employed. While the cutters shown in the illustration are only approximately $1\frac{1}{2}$in. diameter, by virtue of the back gear forming a part of the device it is possible to mount cutters up to $2\frac{1}{2}$in. diameter. When large cutters are set up additional support is sometimes advisable. This takes the form of an overarm that can be fastened to the cross-slide on which the attachment is mounted. The illustration, Fig. 6.26, shows this overarm clearly. Incorporated in it is a threaded centre that can be brought into contact with the end of the cutter arbor and locked when correctly adjusted.

Fig. 6.27   The author's lathe mounted dividing equipment.

In order to cut gears in the lathe some means of indexing the work is needed so that the correct number and spacing of the teeth can be assured. This is the purpose of the dividing head illustrated in Fig. 6.27. It is bolted to a quadrant at the back of the lathe and the worm, with which it is fitted, makes contact with lathe change wheels mounted on an extension to the mandrel. In this way quite extensive dividing can be carried out with relatively simple equipment.

CHAPTER SEVEN

PLANING AND SHAPING MACHINES

IN THE EARLY PART of the 19th century, for lack of more suitable equipment, flat surfaces were machined by a turning operation with the work mounted on a large faceplate in the lathe.

In 1817, Richard Roberts (1789–1864) who has already been recorded as having contributed much to the development of the lathe, designed and built a planing machine which is still in existence in the Natural Sciences Museum, South Kensington, London.

The actual invention of the planing machine, however, is attributed to several engineers. According to one of his own workmen Matthew Murray, who was a competitor of Bolton and Watt in the steam engine field, is said to have invented, if not to have produced, such a machine in 1814.

Joseph Clement, 1779–1844, who had once been in the employment of Bramah, the lock manufacturer, but had then gone into business on his own account, is recorded as having made a planing machine, presumably for his own use, in 1825.

In America, the first planing machine, built about 1836, had a bed made from granite, recessed to take the iron slideways on which it was supported.

But it was not till the Great Exhibition of 1851, when Joseph Whitworth was showing a collection of machine tools far superior to those of his competitors both in design and in scope, that a practical power-driven machine became available. The planing machine in question, illustrated in Fig. 7.1 was made in 1842.

One of the disadvantages of both planing and shaping machines is that in each case, half the operating cycle is idle, that is non-productive. Whitworth endeavoured to overcome this defect by making the tool boxes of his planing machines reversible so that the tools held in them would cut in either direction. He did not, however, have much success so abandoned the idea and concentrated on reducing the idle time to as short a period as possible.

In both planing and shaping machines this reduction can be brought about by an increase in the speed of the return stroke, not a difficult problem when the first-mentioned machine is belt driven.

All that is needed are a pair of fast-and-loose pulleys, generally of the same diameter, driven by belting from two pulleys of different

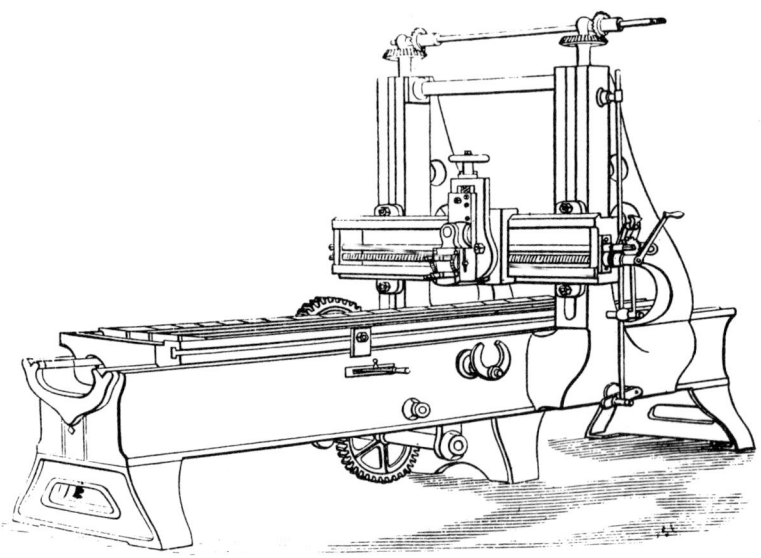

Fig. 7.1   Whitworths Planing Machine.

diameter, the larger being connected to the planing machine by means of a crossed belt so that a reverse stroke of the work table can be made at a higher linear speed than that of the forward stroke.

As to the machine itself, a glance at the illustration will show that it consists of a base provided with two concave V-slideways into which fit the corresponding raised slides machined on the underside of the work table. The mechanism for moving this table comprises a rack affixed to its underside and driven by a pinion attached to a cross shaft fitted with the pulley arrangements described earlier.

The table itself has T-slots and holes so that work may be secured directly by means of bolts.

The tool box assembly is carried on a cross-slide secured in turn to a pair of upright members provided with hand-operated screw gear so that the cross-slide can be adjusted to the correct working height. In order that the cross-slide can be traversed automatically linkage is provided enabling it to do so. This may be seen at the side of the right hand upright behind the traverse handle.

As to the tool-box assembly. This consists of a short slide fitted with a clapper box (whose purpose we shall be describing when discussing

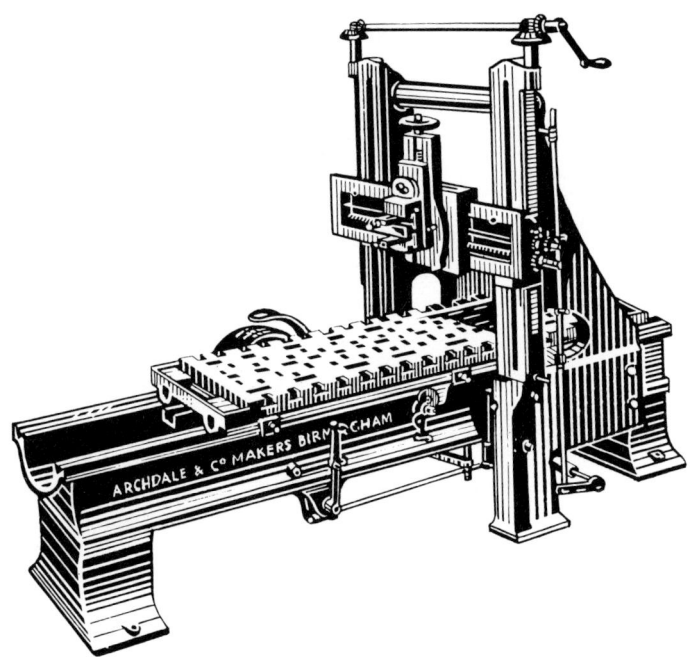

Fig. 7.2   Archdale's Planing Machine.

the shaping machine), and means for clamping the tool itself. The depth of cut is, of course, adjusted by means of the handle attached to the feed screw.

So that it can be reversed automatically it is important to be able to adjust the stroke of the work table. The mechanism for doing so is the horseshoe-shaped object seen in the middle of the base member which is operated by the striker fixed to the side of the work table.

The planing machine depicted in Fig. 7.2 was made by James Archdale of Birmingham in 1884. This illustration shows clearly the arrangement of bolt holes used to secure castings to the worktable, it also demonstrates the placing of the table reverse striking gear.

Fig. 7.3   Open Side Planing Machine.

The Open-Side Machine

Mounting some awkwardly shaped castings in the planing machine were found to be difficult if not impossible to carry out because one or other of the uprights carrying the cross slide got in the way. To overcome this disability the 'open-side' planing machine was introduced, Fig. 7.3.

As its name implies this type of planer has but one support for the cross-slide, a massive column casting supporting the essential elements of the machine and of sufficient rigidity to ensure that no deflection of the cross-slide can take place when the tool-box assembly is furthest away from the column. The open side is one of the latter developments in planing machine design, sometimes incorporating an independently driven milling spindle on the cross-slide as well as the usual tool-box assembly. In addition the work table is now sometimes driven by a hydraulic ram. This enables a wider range of work table speed to be obtained, as well as greatly simplifying the automatic limitation of the table movement.

Securing the Work to the Table

It is of the utmost importance that work is securely fastened to the table, for any 'spring' that can take place will undoubtedly result in

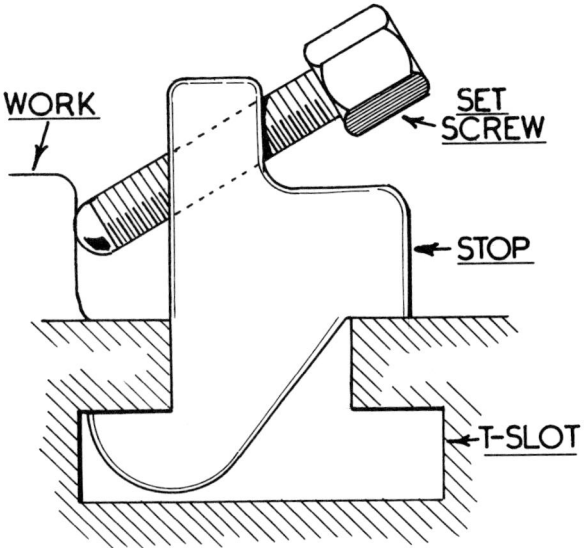

Fig. 7.4　Stop for securing work.

inaccurate machining. Bolts, clamps and dogs of one sort or another are used for the purpose, the design of castings being such as to assist in this matter. In addition a special solid form of stop such as that illustrated in Fig. 7.4, is sometimes used to reduce spring at any point where it is found likely to occur.

These stops, one of which is illustrated, are designed to fit in the table T-slots. They are provided with set screws so that downward pressure can be applied to the casting ensuring that it beds on the table firmly. They are used in conjunction with shims, where necessary, placed under the work to afford support and make certain that the stop itself does not distort it.

The Shaping Machine

As a tool for use in the jobbing shop, where the production of small batches and even single components is the general rule, the shaper has many advantages over the milling machine. No special cutters are needed, since the tools used with the lathe can generally be employed.

For the most part the shaping machine is used on small components where it would be uneconomical to employ the planer.

The shaping machine is made in two forms. In the first of these the work table is held stationary whilst the ram carrying the tool travels along the horizontal slideway allowing the work to be fully covered by the machining operation.

Fig. 7.5   Travelling Head Shaper.

The second, and now perhaps the most commonly used type of machine, employs a reversal of this arrangement with the work itself travelling along on a table supported by a horizontal slideway forming part of the main frame of the machine. In this case the tool ram is carried in a fixed slide machined in the upper part of the principal casting.

Both classes of machine are illustrated, the first in Fig. 7.5, the second in Fig. 7.6.

Fig. 7.6   Fixed Head Shaper.

Generally, the principle upon which they operate is the same, the cutting tool being carried in a box attached to a slide secured to a power-driven ram while the work is secured to a table so that the tool itself can be made to pass over the surface of the work either by hand or automatically.

Planing, Shaping and Slotting Machines

The internal mechanism of the type of shaper illustrated in Fig. 7.6 is depicted in Fig. 7.7. The crank arrangements are different from those that Whitworth knew, but the disc crank is still in evidence and there is, of course, provision for adjusting the stroke of the ram as well as its position in relation to the work itself. The particular mechanism illustrated is that of the G.S. range of Butler shapers. Here, the helical-toothed stroke wheel 'A' carries a cast-steel slider 'B' provided with a large diameter driving pin. The slider itself runs in dovetails formed in

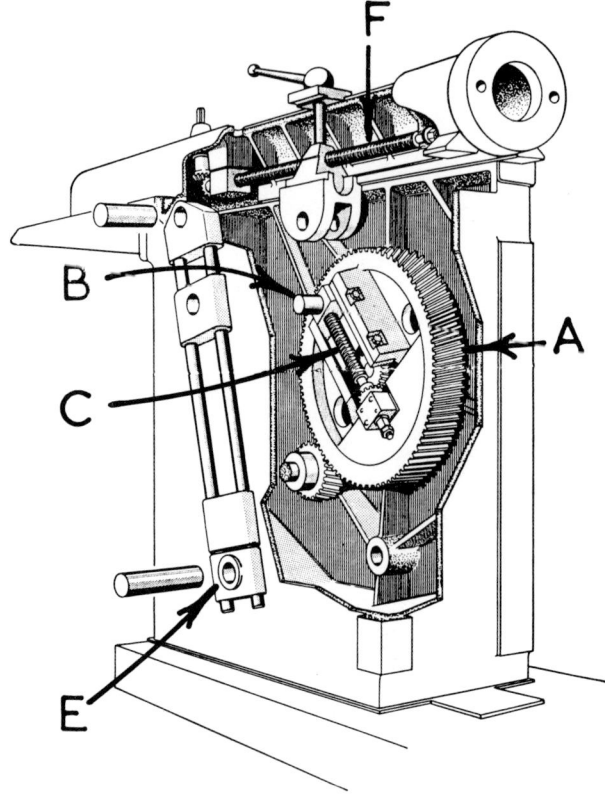

Fig. 7.7   Internal Mechanism of Butler Shaper.

the body of the wheel, being moved by a screw 'C' that may be operated from outside the shaper, through the bevel gears seen in the illustration.

Two steel rods pivoted in the ram, pass through, a driving block 'D' that engages the pin forming part of the cast-steel slider. The two steel rods are supported by a trunnion bearing 'E' that pivots around the pin seen at the bottom of the illustration.

The position of the ram in relation to the work is adjusted by the screw 'F' operated by a key through a pair of bevel gears. The screw passes through the ram pivot which is locked and unlocked by the lever seen at the top of the illustration.

A six-speed gearbox takes the drive to the pinion in mesh with the stroke wheel. Automatic cross feed is variable from 0in. to 0·060in.

Certain requirements are essential to the shaping machine. First of all it must be possible to so adjust the position of the ram in relation to the work that the work surface itself is fully covered by the ram's movements. Secondly the length of the stroke must be capable of adjustment so that idle time can be avoided or at least cut as short as may be. The third and last requirement is the provision of mechanism enabling the work to be traversed under the tool automatically during the machining of extensive surfaces. The traverse mechanism itself has also to be capable of adjustment, in order that the operator may make an alteration in the feed rate when needed; for example when taking a roughing cut over a surface where fine finish is not required.

Joseph Whitworth was fully aware of these criteria and in his shaper, the elements of which are depicted in the illustration Fig. 7.8, he incorporated them. In addition, but adjusting the geometry of his crankshaft in relation to the axis of the ram itself, he was able to provide for its quick return on the idle or non-cutting stroke.

As will be seen the disc crank, providing movement to the connecting rod, is driven from a pinion mounted on a shaft running along the back of the shaping machine, the crank itself forming part of a gear wheel assembly with which the pinion is in mesh. The crank-pin is

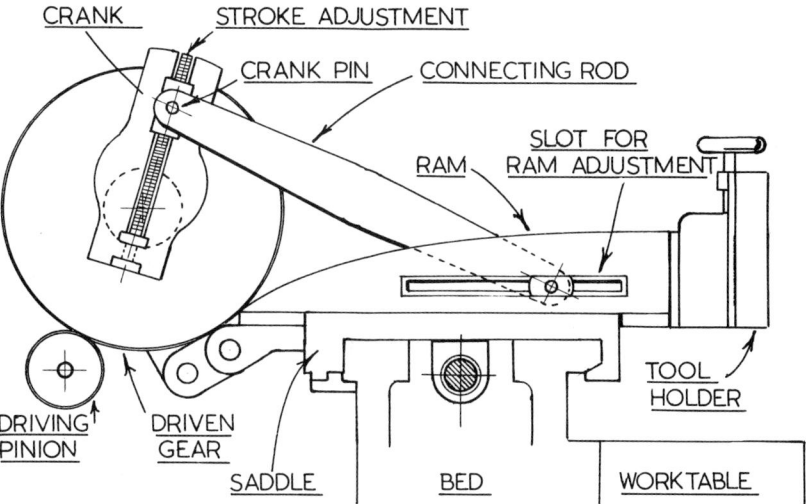

Fig. 7.8   Diagram of Whitworth's Shaping Machine.

E

capable of positional adjustment in relation to the centre of the disc crank, a mechanism being provided for the purpose. So the stroke of the ram can be set to any length within the capacity of the machine.

Self-acting mechanism was also fitted to the traverse screw, though this does not appear in the illustration.

A common method of providing automatic traversing is by means of a ratchet wheel and reversible pawl, the ratchet being fitted to the feed screw, whilst the pawl is operated by a connecting rod attached to a disc crank driven through gearing from the main drive shaft. The elements of the device are illustrated in Fig. 7.9.

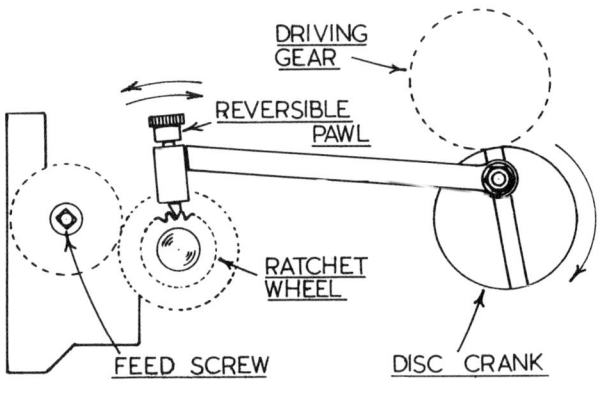

Fig. 7.9   The Self-act.

The disc crank mechanism is capable of adjustment enabling the crank-pin setting itself to be altered in order to vary the stroke of the ratchet and so produce a coarse or fine feed as and when desired.

This is the arrangement generally applied to small power-driven machines, and is the one to be found on the shaping machine in the author's workshop. Large machines, however, are often fitted with a gearbox to provide a more varied and rapidly adjustable rate of feed.

Shaping Machine Tools

Whilst tools used in the lathe can be transferred to the shaping machine and employed successfully, though in the main for light cuts, when heavy machining is contemplated, however, a swan-neck tool of the type depicted in Fig. 7.10 is better applied to the work. In a light machine the theoretically ideal tool is one having its cutting edge behind a line passing through the point of support as shown in Fig. 7.11 at B. Here the point of the tool, if deflected, describes an arc AB

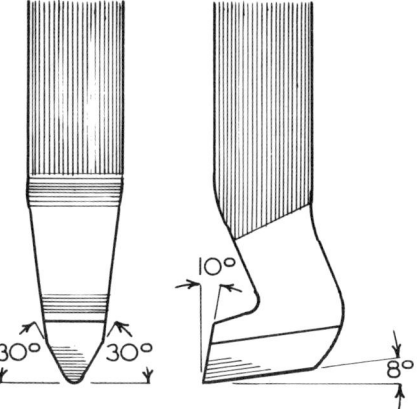

Fig. 7.10   Swan-neck shaping machine tool.

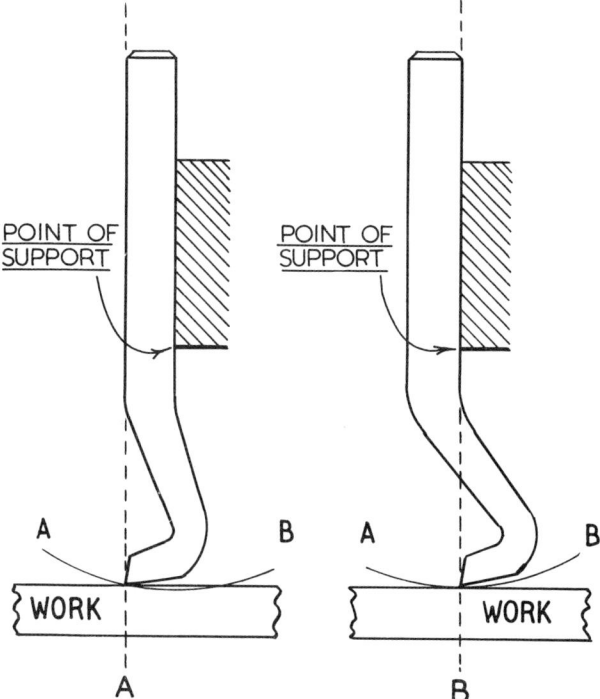

Fig. 7.11   Theoretical disposition of shaping machine tool.

that tends to clear the work increasingly as the load is applied. Not so the tool illustrated in Fig. 7.11 at A; in this case the tool point is on an arc that will cause it to bite deeper into the work. The tools shown in the illustration are for diagrammatic purposes only. Were they used in practice, their inherent weakness would cause them to 'unwind' thus producing a condition even worse than the one the second tool was intended to overcome.

Mounting the Tool

In the development of the shaping machine, as with the planing machine, it was realised early that the cutting tool needed to be relieved on its return stroke; the more so because the wearing qualities of the available tool steel were poor.

The device used to provide the relief is called the 'clapper box'. This is a metal block hinged at its upper end and carried in the tool slide attached to the ram. On the forward stroke while the tool is cutting, the block abuts against its seating, while on the reverse stroke the block hinges away from the seating allowing the cutting edge of the tool to lift slightly and be relieved of pressure.

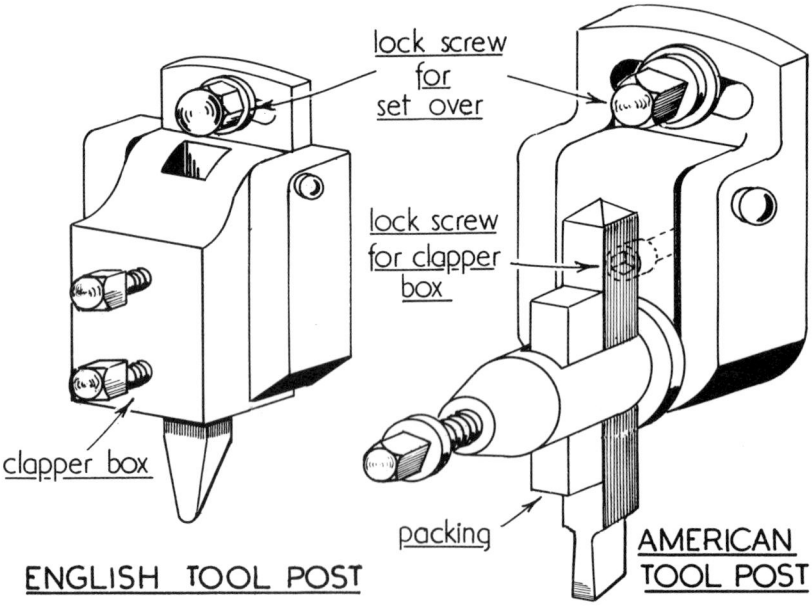

Fig. 7.12   English and American Toolposts.

Fig. 7.13   American Toolpost fitted to a shaping machine.

The tool itself can be mounted either directly in or on the clapper box as was, generally, early English practice; or it can be set in a tool post of lantern or American pattern attached to the clapper box. Both arrangements are illustrated in Fig. 7.12, while a practical example of the second type, attached to a shaper in the authors workshop, is seen in Fig. 7.13.

It is sometimes desirable to set over the clapper-box assembly when machining some components. The illustration Fig. 7.12 shows that provision is made for this, but, in the case of the English toolpost, the amount of set-over is limited. With the American toolpost, however, the range is greatly increased because the post itself, as has been described in Chapter 3, can be swung through 360 degrees.

Mounting the Work

Small components are usually gripped in a machine vice bolted on the work table. The vice itself is secured to a base around which it may be

Fig. 7.14   Slotting machine by Butler of Halifax.

rotated and then locked when machining at a predetermined angle has to be undertaken.

Large parts are bolted directly to the work table, usually a box-form casting, upon which the work may be set either on its upper surface or secured to either side as is most convenient at the time.

The Slotting Machine

The mechanism of the modern slotting machine is basically the same as that of the shaper, with the ram vertical instead of horizontal. This

Fig. 7.15 Slotter with outside counterweight.

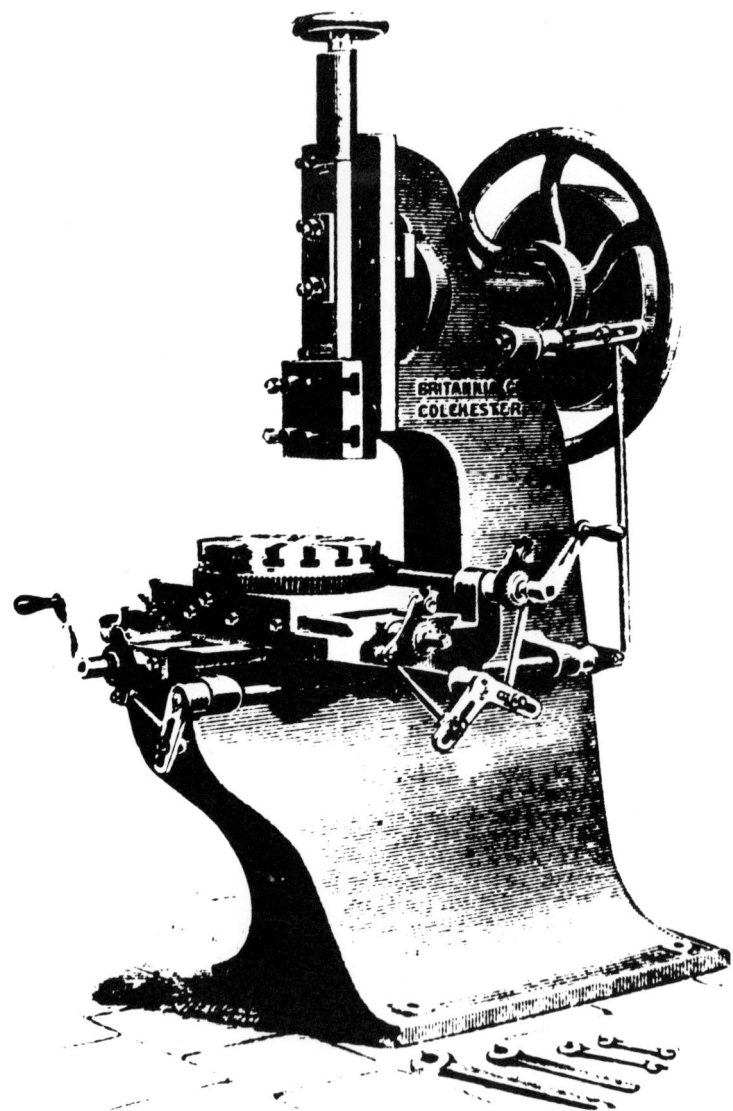

Fig. 7.16   Slotter of 1880.

configuration enables the slotter to be used for such work as the cutting of internal keyways, the radial shaping of selected components using a rotary table, and any work needing indexing.

Slotting machines date from the same period as the shaper, and have developed in much the same way; till, now, they have built-in electric motor drive, a range of speeds derived from a gearbox, and a rotary table that may be power-operated when required. A typical example by Butler's of Halifax is illustrated in Fig. 7.14.

Unlike the shaping machine the ram of the slotter needs a counter-balance. In a modern design the counter-balancing weight would be contained inside the head casting, as opposed to early machines when the weight pivoted in lugs cast on the top of the head itself. The arrangement being similar to that shown in the illustration, Fig. 7.15, which depicts a slotting machine made about 1910. On the other hand a slotter, Fig. 7.16, made sometime previous to 1880, and taken from an illustration in the Encyclopaedia Brittanica of that date, would seem to indicate that the earliest machines, probably owing their origin to Richard Roberts of Manchester, had no counterpoise at all.

It is said, and this seems very likely, that Roberts obtained the idea from the wood morticing machine which was of course hand-operated but possessed all the basic movements essential to the slotting machine.

CHAPTER EIGHT

THE DRILLING MACHINE

NEXT TO THE LATHE the drilling machine is probably the oldest of man's mechanical tools. It is reputed to have existed in the year 1600, but was probably then no more than a method of securing and applying pressure to a carpenter's hand brace in the manner depicted by the illustration Fig. 8.1. This very elementary arrangement, which must have been commonly employed because it again appears in an illustration published at the latter part of the 18th century and reproduced in Fig. 8.2, was later developed into the more advanced device illustrated in Fig. 8.3. The column 'a' is caught in a leg vice attached to a three-legged stool while a cross bar 'b' secured to the column by means of the screw 'C'. The hand brace is retained in place by the screw 'e' which is also used to apply the pressure needed to make the drill cut. Drilling equipment of this nature probably dates from about 1700 and may well have served for the many purposes requiring little accuracy.

Jacques de Vaucanson, 1709–1802, of whom we have heard in connection with the milling machine, was also the developer of a more sophisticated drilling machine described as having the height of the drill itself controlled by one leadscrew whilst the horizontal position of the drill point was established by a leadscrew operating a carriage running on a pair of prismatic-shaped rails. The drill spindle itself was apparently driven by round belting from overhead shafting. The date of the introduction of this machine is uncertain. However, if we assume it to be about 1750 then power driven drilling machines have been in use for some 200 years. By the end of the eighteenth century considerable advance had been made in general tooling and some of these improvements brushed off on to the drilling machine itself. One example of the advance being made is the portable hand drill of James Nasmyth 1808–1890. This machine was introduced about 1840 and is of special interest because here we have the first instance of a drilling machine with bevel gear drive and a method of feeding the drill that, with improvements and some variants, was to remain as standard for a long while. The machine itself is illustrated in Fig. 8.4.

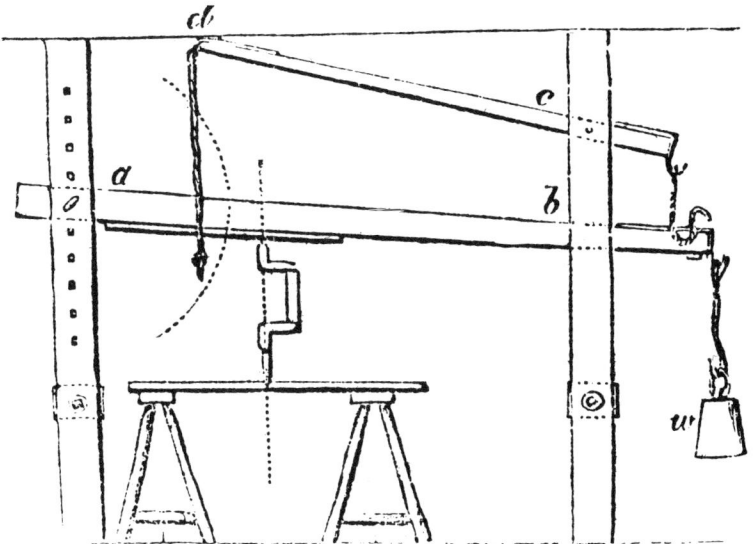

Fig. 8.1 Early methods of applying pressure to the Carpenters Handbrace.

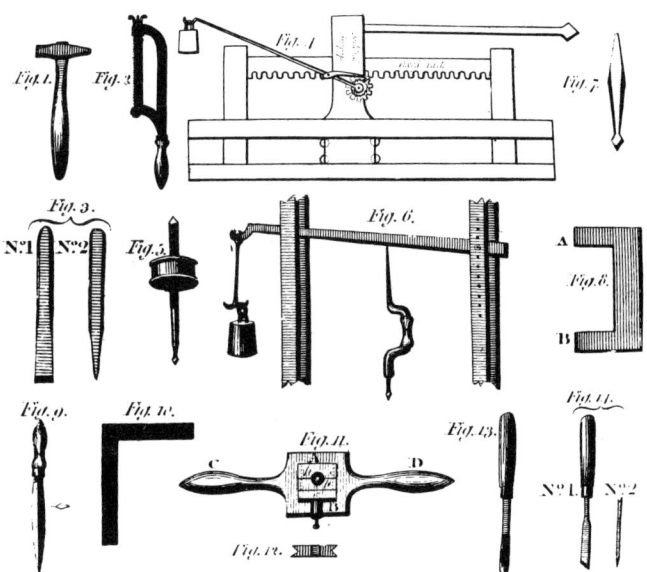

Fig. 8.2 Early methods of applying pressure to the Carpenters Handbrace.

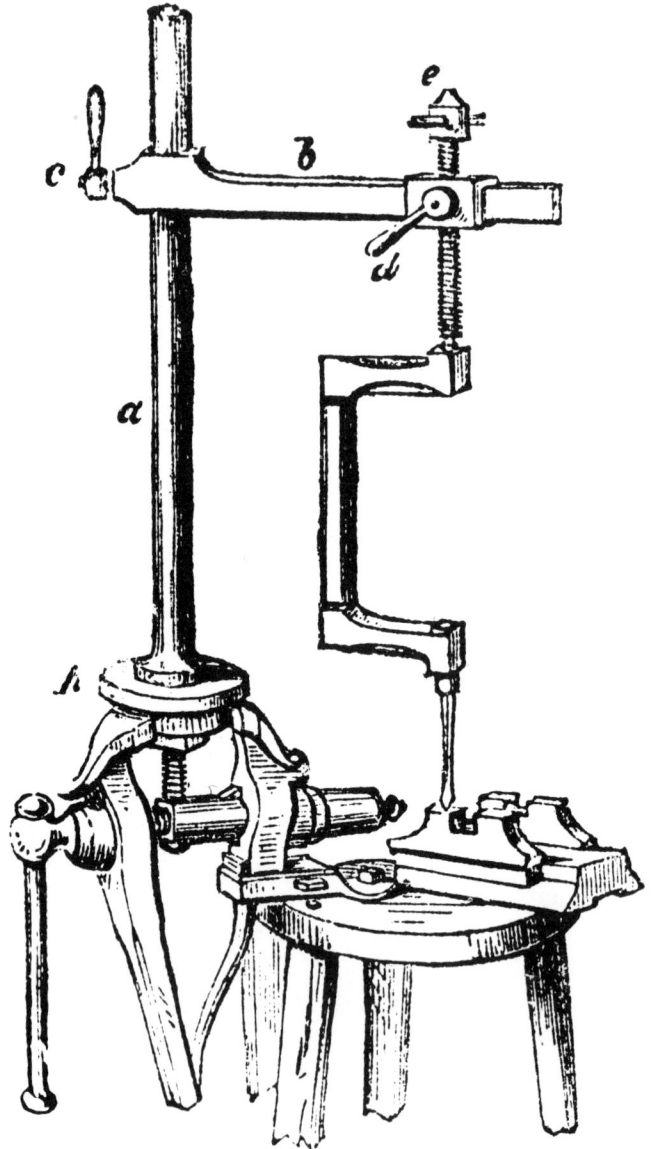

Fig. 8.3   Blacksmith's Drilling device.

Fig. 8.4   Nasmyth's Hand drill.

We have already seen the contribution to machine tool development being made by Sir Joseph Whitworth, 1803–1887, and how he was leaving his competitors far behind both in the quantity and quality, as well as in the wide range of machine tools produced by him. It is not surprising, therefore, to learn that his firm, Whitworth & Company of Manchester, had developed a power driven drilling machine in 1847 that was complete in every detail, having self-act to the drill feed, a back gear making a range of spindle speeds available, and lastly a rotary table with lateral movement to assist in aligning the work with the drill point the more readily.

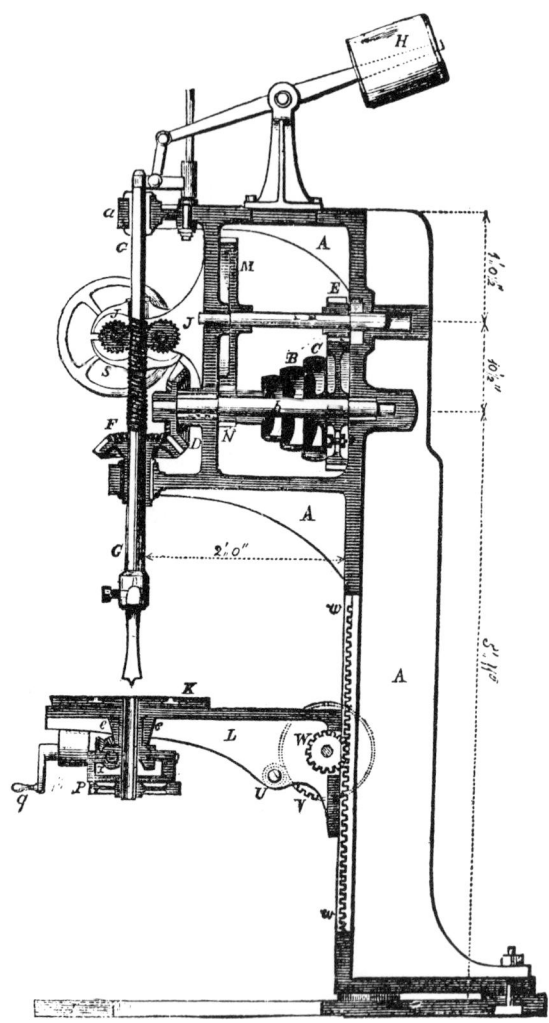

Fig. 8.5　Whitworths Drilling machine.

This machine is illustrated in Fig. 8.5. In his 'Encyclopaedia of Useful Arts' 1854 Tomlinson remarks:

"Messrs. Whitworth & Co. of Manchester have constructed a drilling machine which is one of the most complete tools of the kind ever constructed".

During the construction of the buildings that housed the Great Exhibition of 1851 a simple single speed drilling machine was in use for drilling the various ironwork components needed in the building. This machine, which is depicted in Fig. 8.6, is of interest because the drill feed was applied by the foot, leaving both the workman's hands free to align the work and hold it in place.

Fig. 8.6 Single-speed drilling machine used on the construction of the Great Exhibition 1851.

In 1860 a paper was read to the Institute of Mechanical Engineers describing the equipment that was used to drill the girders of the Hungerford railway bridge at Charing Cross in London. This work involved the drilling of thousands of rivet holes in the shortest possible time. The machine used was designed to drill 1in. diameter holes at a feed rate of 0·040in. per minute. It had 20 spindles and the worktable was raised hydraulically to provide the feed. This was probably the first example of simultaneous multi-spindle drilling, a practice now much employed industrially.

By 1884, a typical English manufacturer James Archdale & Company of Birmingham were cataloguing a range of drilling machines including a non-elevating radial arm drill illustrated in Fig. 8.7, a centralised control Radial Drilling Machine, Fig. 8.8, and the double geared vertical drill seen in Fig. 8.9.

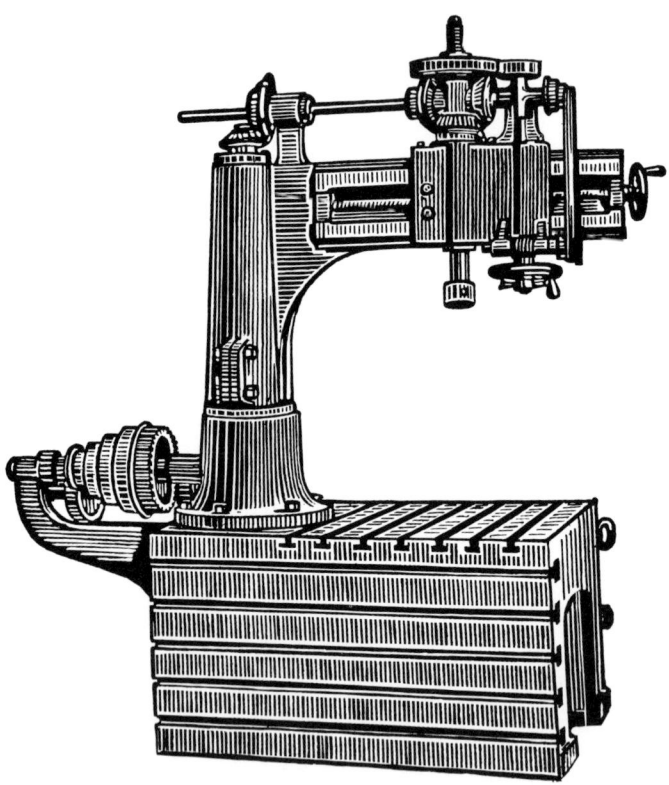

Fig. 8.7   James Archdale & Co's. non-elevating radial arm drill.

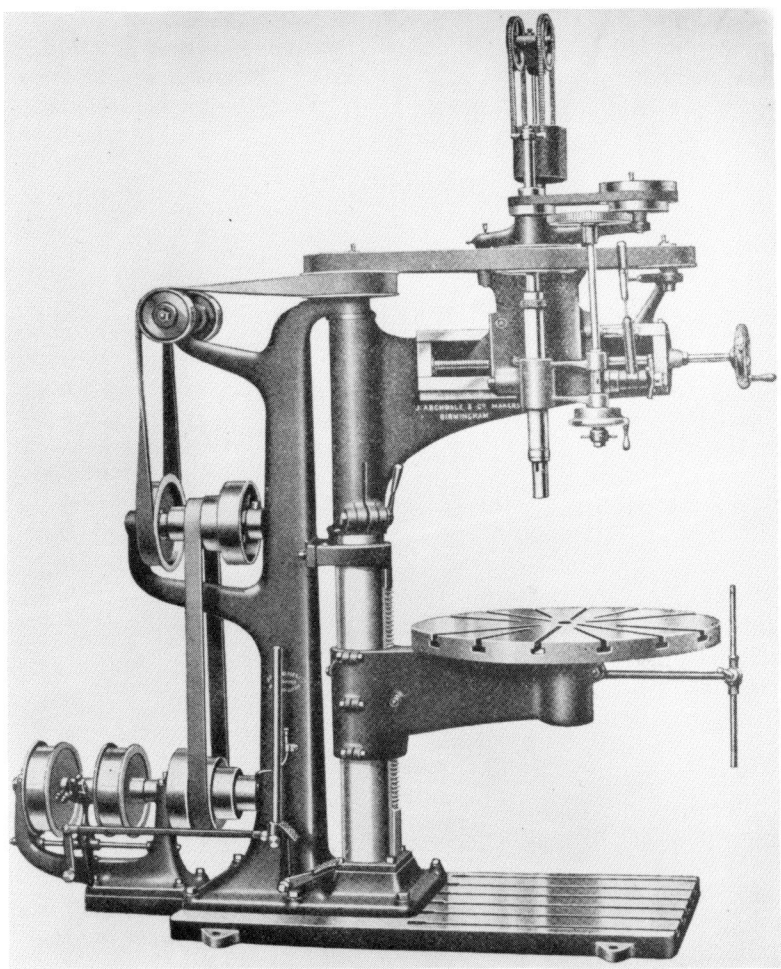

Fig. 8.8 Archdale Centralised Control Radial Drilling Machine, 1897.

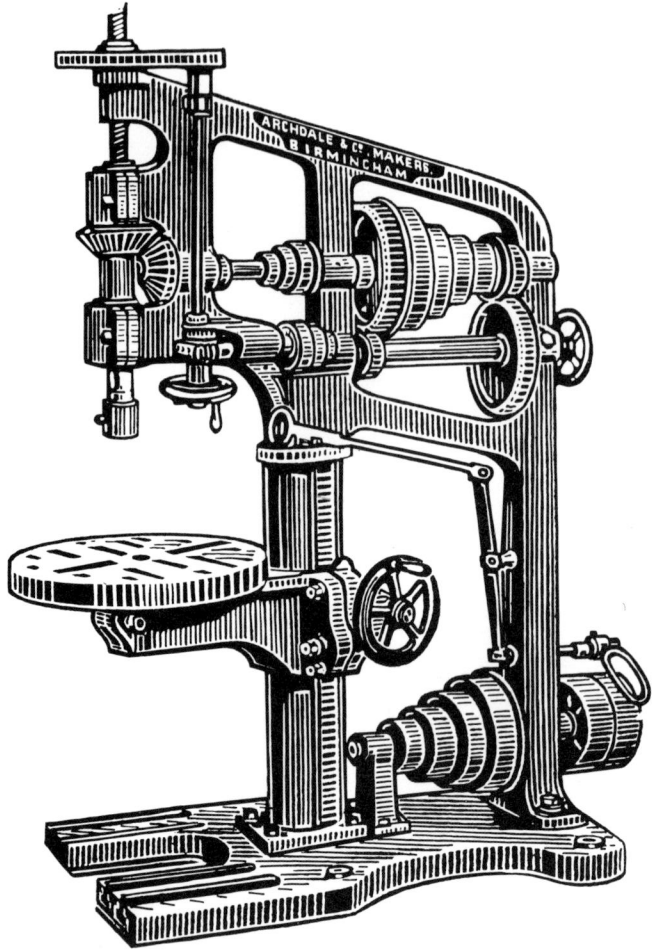

Fig. 8.9   Archdale Double-Geared Vertical Drill, 1884.

Methods of Holding the Drill

Before we go on to consider the various types of drill that have been, and in some instances still are available to the user of the drilling machine, some words ought to be said about the various methods that have been employed to mount the drill in the machine spindle.

Early machines, having had the carpenter's hand brace for parentage, understandably made use of the arrangement obtaining in that tool.

Fig. 8.10 demonstrates the method employed, and it will be seen that the taper squared shank of the drill if fitted into a corresponding recess in the spindle of the drilling machine itself, a set screw, for the most part engaging a dimple in its shank, holding the drill in place.

Where little accuracy was expected, however, such an arrangement was well enough. But the advent of the twist drill from America in the latter end of the nineteenth century, with its far greater degree of accuracy, entailed the employment of the American system of drill mounting in order to preserve this accuracy.

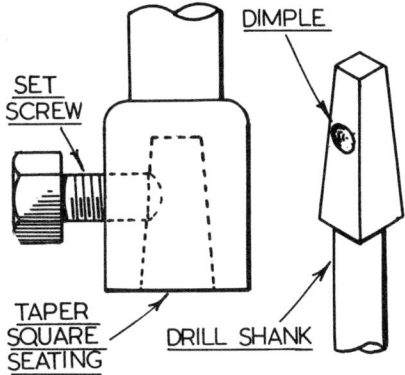

Fig. 8.10   Method of holding the drill in early machines.

The American Morse Twist Drill Company had produced a drill with a comparatively long slow-tapered shank which, when placed in a mating female taper machined in the drill spindle provided sufficient grip, for the most part, to drive the drill and ensure that it ran truly.

However, to guard against the possibility of the drill turning in the spindle as well as to provide a simple way of withdrawing it when necessary, a tang or key was formed at the top of the drill shank and the drilling machine spindle given a recess to receive it.

The set-up is depicted in Fig. 8.11 where the drift used to remove the drill and the tang against which it operates are illustrated.

Such an arrangement was applied to the larger sizes of drill for the most part, the Morse Tapers as they were called—and they have since become a standard for drill purposes—increased in size as did the drills themselves. But for the smaller drills, and even for those commonly used drills for say $\frac{1}{4}$in. diameter to $\frac{1}{2}$in. diameter, the provision of a tapered shank for each size would have been wasteful though sometimes, industrially, this extra cost is justifiable.

Accordingly, several forms of chuck were tried in attempts to hold these smaller drills accurately. Some were partially successful, but many did actual damage to really small drills as the author can personally testify. So it was not till 1903 when, in October of that year, A. I. Jacobs of the Jacobs Manufacturing Company, Hartford, Connecticut, U.S.A., evolved a unique form of chuck that has since

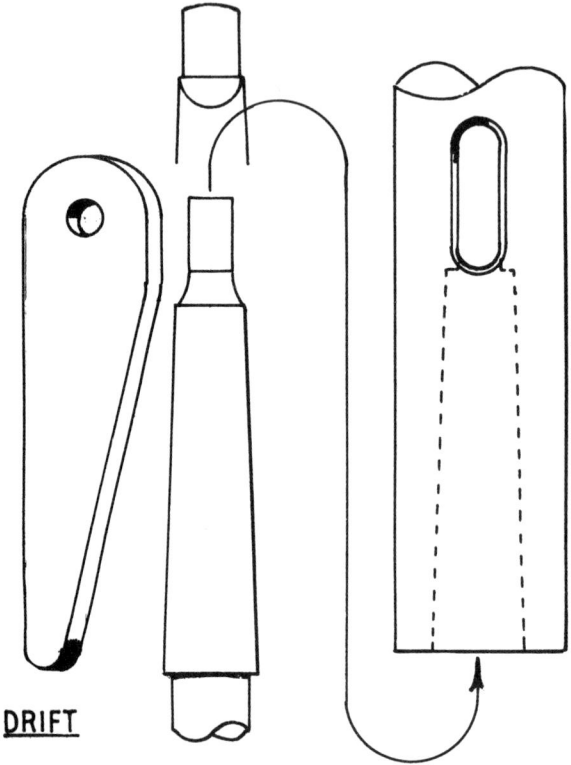

DRIFT

Fig. 8.11   Morse Taper Shank drill mounting.

become virtually the standard for holding drills of many sizes, each chuck for the most part being able to grip accurately drills of zero diameter up to the maximum that the chuck will hold. The Jacobs chuck is mounted in a drilling machine either by being set on a tapered arbor, similar to that of the taper shank drill, (a chuck for this purpose is illustrated in Fig. 8.12), or it may be screwed directly to the machine spindle, a chuck suitable for this duty being depicted in Fig. 8.13. Examples of this last application being the many electric hand drills now on the market.

The sectioned chucks show clearly how they are constructed. The body 'A' houses three jaws 'B' free to slide in housings formed in the body. As these housings are set at an angle it follows that, when the jaws move simultaneously under the control of the nut 'C' they can be made to close on any drill placed within them. The nut is turned by the sleeve 'D' in which it is friction-tight, the sleeve having gear teeth

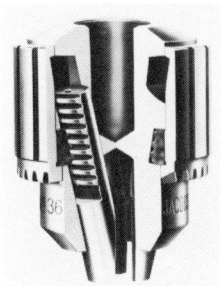 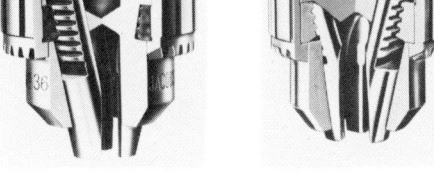 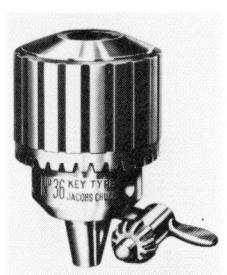

Fig. 8.12 Jacobs chuck for taper mounting.

Fig. 8.13 Jacobs chuck for screw mounting.

Fig. 8.14 Jacobs chuck and key.

machined on it as seen in Fig. 8.14. Also depicted in this illustration is the key supplied with the chuck. This, too, has corresponding teeth and is inserted in one of three holes in the body when the chuck needs to be tightened.

Drills

The drills once used were of simple construction and, for the most part, readily shapened by the rudimentary equipment then available.

The original drill for metal is that illustrated in Fig. 8.15. This is the spearpoint drill that must have been used universally until the advent of the straight fluted drill. The spearpoint drill could be produced by a forging process so was easily made and hardened by the local smith. It can have had but little accuracy and was probably given to 'wandering', a fault that may have given rise later to the provision of a pilot at the apex of the drill point as seen in Fig. 8.16 in an attempt to minimise the trouble.

The requirements of industry could not long tolerate the inaccuracies of these blacksmithian devices, so it was not long before the demand for drills of proven accuracy began to be made. No doubt as a direct result of the requirements for armaments resulting from their civil war, the Americans produced first of all the straight flute drill already mentioned, and then the twist drill which is now the accepted

Fig. 8.15 The spear-point drill.

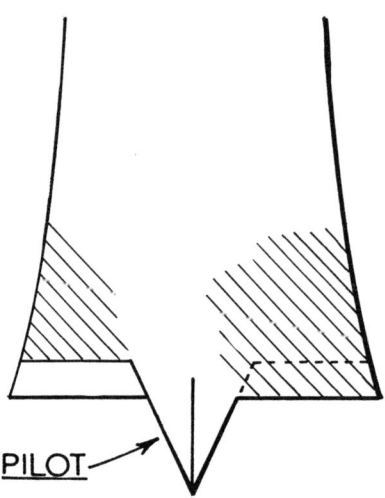

PILOT

Fig. 8.16 Modified spear-point drill.

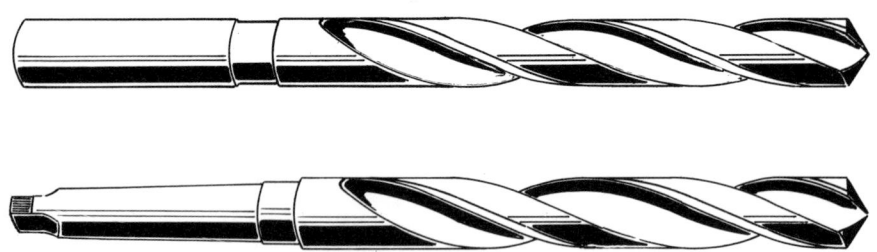

Fig. 8.17 Shank Twist Drills Parallel Shank and Taper.

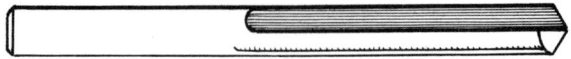

Fig. 8.18 Straight Flute Drill.

standard type. By adopting careful methods of manufacture it was possible to produce these drills in quantity and to control their sizes closely.

The straight fluted drill is seen in Fig. 8.18. This type of drill is admirably suited to working in brass as it does not 'grab' and it is said that at one time Rolls-Royce or maybe Sir Henry Royce himself would allow no other in their brass-finishing shops!

One disadvantage of the straight-flute drill, particularly in deep drilling, is its inability to clear swarf quickly. It is also somewhat slow in cutting steel, a fault that led to the introduction of the spiral-fluted or twist drill illustrated in Fig. 8.17. The particular conformation of the drill itself presents a better cutting angle to the work and greatly assists

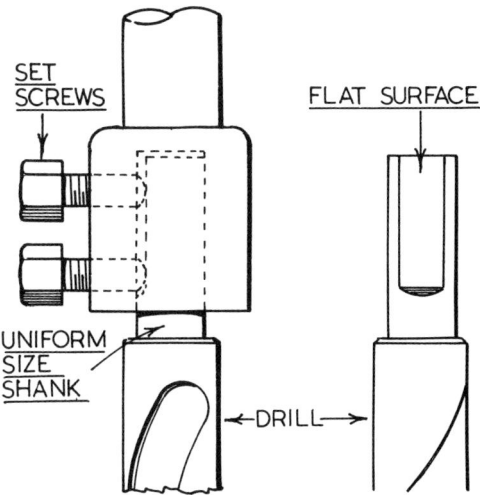

Fig. 8.19   Alternative direct method of mounting parallel shank drills.

in swarf clearance. Two forms of drill are shown in the illustration. The parallel shank for use with a chuck, and a taper shank for direct mounting in the machine spindle, a method that has been explained earlier. An alternative method of mounting the parallel shank drill allows it also to be set directly in the spindle. This, as depicted in Fig. 8.19, involves the provision of a uniform size parallel shank having a flat surface machined upon it. This surface acts as an abutment for one or more set screws located in the drill spindle itself.

Any accuracy that this arrangement possesses clearly depends on the closeness of the fit between the shank of the drill and the corresponding socket in the drill spindle.

Over the years the pitch of the drill fluting has been varied to suit the work in hand as the requirements of industry dictated. Fig. 8.20 depicts some of the pitch variants available.

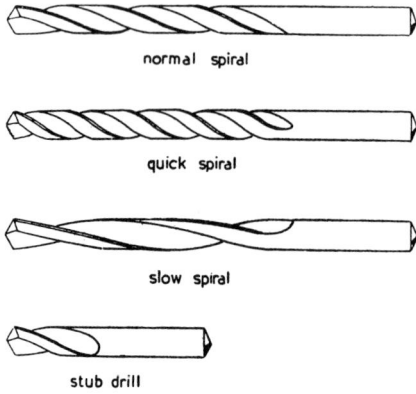

normal spiral

quick spiral

slow spiral

stub drill

Fig. 8.20   Twist drill pitch.

Bits for Drilling Wood

As may be expected the drilling machine is not confined to handling metal components. In common with other machine tools it has a large part to play in the drilling of wooden parts. For the purpose, the drill bits once used for the hand brace are capable of employment in a machine though not necessarily with any certainty of accuracy. The

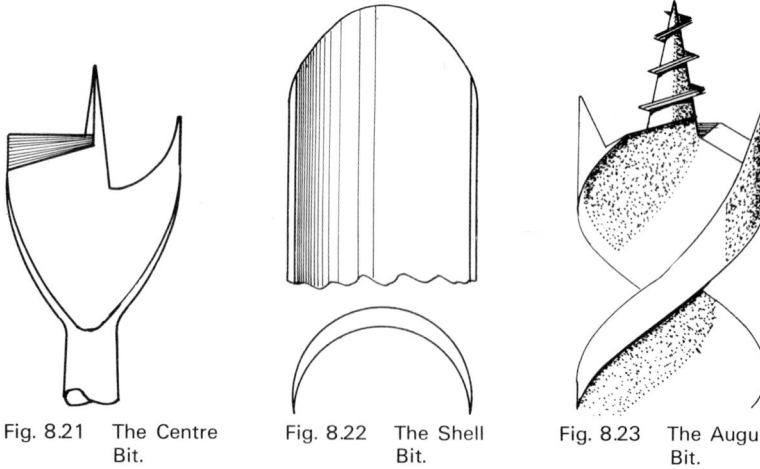

Fig. 8.21   The Centre
Bit.

Fig. 8.22   The Shell
Bit.

Fig. 8.23   The Augur
Bit.

centre bit, Fig. 8.21, must at one time have been used for shallow drilling, indeed the author has sometimes employed such a drill for this purpose in a machine. It is essential, however, to modify the shank to ensure that the point runs truly. The shell bit, Fig. 8.22, was intended for deep drilling but, if used in a machine, would need frequent retraction to clear wood chips. The auger bit, Fig. 8.23, is the best tool for both shallow and deep drilling as it cuts cleanly, maintains reason-

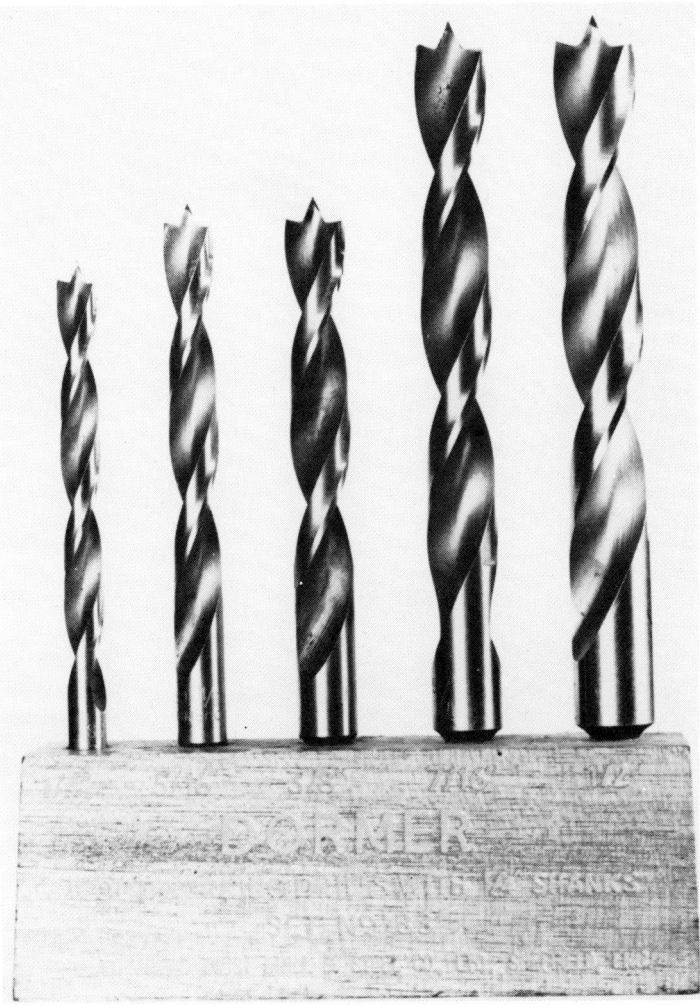

Fig. 8.24   Modern Drills for wood.

able accuracy, and ejects the wood chips rapidly. But, if used in a drilling machine, the pilot needs to be made smooth or the auger bit will take charge and drag the feed lever out of the operators hand. The logical development from the common auger are the wood drills illustrated in Fig. 8.24. These are a product of the Dormer Drill Company, and are in reality, modified twist drills. They have accurately centred pilots and a pair of spurs that cut cleanly to close limits dimensionally. At the turn of the century a wide range of drills of this type was available. Now apparently only the few seen in the illustration are to be had, making a set of five drills from $\frac{1}{4}$in. to $\frac{1}{2}$in. increasing by $\frac{1}{16}$in.

The Countersink and Counterbore

In order to seat the heads of some screws correctly it is necessary to machine away some of the material around the entry of the hole that has already been drilled to receive the screw itself. For the most part, two forms of screwhead, are concerned; these are the countersunk head and the cheese head illustrated in Fig. 8.25.

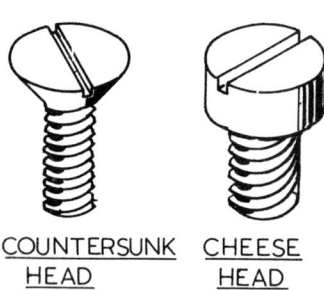

COUNTERSUNK   CHEESE
HEAD   HEAD

Fig. 8.25   Countersunk and Cheese
head screws.

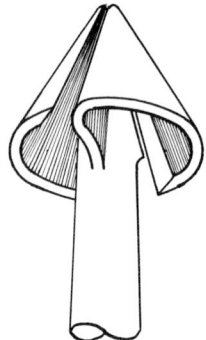

Fig. 8.26   The Wood
countersink.

It is important that the seatings for both types of screw head should be accurately formed and should be without 'chatter' or the appearance of work may be impaired. In the case of wooden components the countersunk, illustrated in Fig. 8.26, was developed early this century by the L. S. Starret Company of America, it is of 'snail' form and has two cutting edges. It is possible to keep this tool really sharp, an absolute necessity of the countersinking is to be well finished.

The original countersinks for metal were flat, resembling the spear-point drill already illustrated. These tools can hardly have been expected to produce a smooth hole and it was not until multi-tooth cutters became available that good results were obtainable from the countersinking process. A typical countersink is depicted in Fig. 8.27 at A. The results from it are not always of the best, the more so when the cutter is run too fast. The countersink seen at 'B', and named the

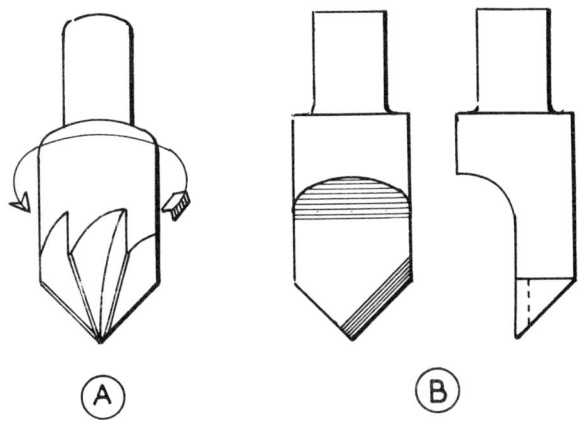

Fig. 8.27 The Metal Countersink.

'accracount', has been developed to overcome any tendency to 'chatter'. It has but one cutting edge lying on the centre line of the body, and this is supported by the plain portion making contact with the work surface. The author can personally testify to the benefits that result from this device.

Spotfacing

The object of spotfacing around a drilled hole is to provide a good seating for nuts or cheese-head screws. This is especially needed when

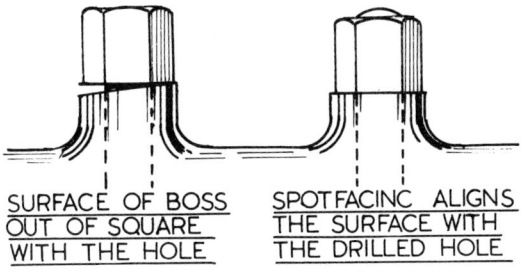

SURFACE OF BOSS OUT OF SQUARE WITH THE HOLE   SPOTFACINC ALIGNS THE SURFACE WITH THE DRILLED HOLE

Fig. 8.28 Spotfacing.

seating them on cast surfaces, though sometimes the spotfacing is dispensed with here to the detriment of the work finish. These alternatives are seen in Fig. 8.28.

The tool used for the purpose is sometimes called a 'pin drill', a name derived from the pin that forms an essential part of its make-up. This pin is made a close running fit in the drilled hole and serves to retain the cutter in place during the spotfacing operation. Early cutters had but a pair of cutting edges and this resembled the pin drills that amateur workers often make for themselves. An example of such a tool is illustrated in Fig. 8.29.

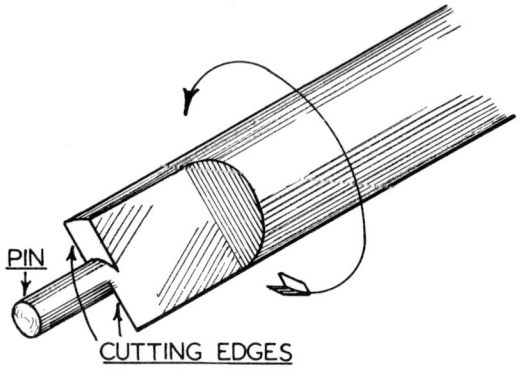

Fig. 8.29   Pin drill or spot face cutter.

As will be seen the pin here is integral with the body as a whole, this practice applying, for the most part, to the majority of commercially made spot facing cutters. However, it is possible to obtain cutters in which the pin or pilot as it is generally known, is detachable and can be interchanged with others of differing sizes. Such an arrangement provides for some increase in flexibility and range combined with economy in production.

CHAPTER NINE

METAL FORGING EQUIPMENT

ONE OF THE EARLIEST, if not the earliest machines to be mechanised was the hammer. The Chinese, some 2,000 years ago, seem to have been the first to do so while the German writer Agricola in his book 'De le Metallica' published in 1556 describes a power hammer driven by a water wheel having the haft, which was pivoted at one end, lifted by cams affixed to a drum mounted on the water wheel shaft in the manner depicted by the illustration Fig. 9.1.

Mechanical hammers of this type are known as lift hammers. The alternative method was to pivot the hammer haft well behind its centre point and to tilt the haft by means of the cam wheel as illustrated in Fig. 9.2, allowing the hammer head to fall on the work when the end of the haft cleared the cams. Devices of this nature are known as tilt hammers.

Hammer mills driven by water wheels seem to have been in general use from the middle ages onwards, and even as late at 1942 a pair of tilt hammers were operating near Newton Abbot in Devonshire, being used for the forging of bill hooks, fisherman's knives and implements for agricultural purposes.

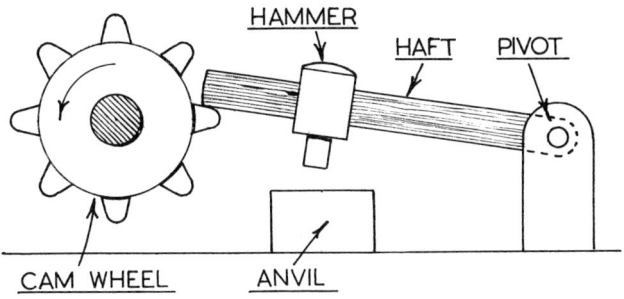

Fig. 9.1 The lift hammer.

127

The firm of Boulton and Watt harnessed their Rotative Steam Engine to the hammers setting up a combined lift and tilt hammer mill for Wilkinson at Bradley Forge in 1783. This was the first practical application of steam to the process of metal forging.

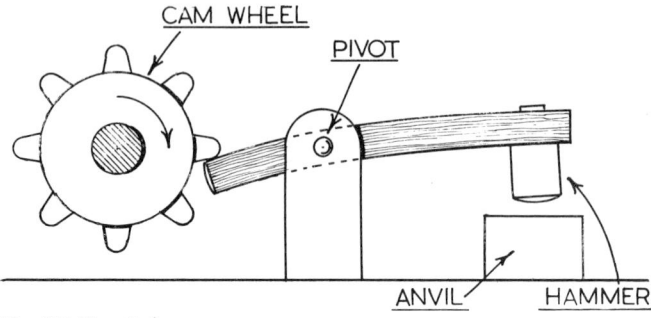

Fig. 9.2 The tilt hammer.

The Steam Hammer

By the middle of the 19th century the hammer mill was gradually being superseded by the steam hammer which had great flexibility in use and vastly increased power. James Watt himself had, in 1756, suggested the direct use of steam but did not make a practical trial of his own suggestion. In 1806 one Deveral drew up plans for a steam hammer, but there appears to be no record of it having ever been made.

It remained, therefore, for James Nasmyth, in 1842, to make the first working steam hammer. Nasmyth's hammer is illustrated in Fig. 9.3. As has been said it was able to deal a far heavier blow than the old lift and tilt hammers whose hammer heads for the most part only weighed from one to five hundredweight.

Originally the steam was applied to the underside of the working piston only, the hammer being allowed to fall by its own weight; but later power was applied to the head of the piston, greatly accelerating the rate of the hammer's fall. As a result the blow given was enormously increased.

The requirements of heavy industry for capability in forging the large components needed in the armament and marine fields were soon on the way to being satisfied by the steam hammer, for the size of these machines was gradually increasing until, by 1865, a hammer set up in Sheffield was capable of delivering a 25-ton blow.

The author well remembers, as a boy, the two large steam hammers once installed in Portsmouth Dockyard and Woolwich Arsenal. When

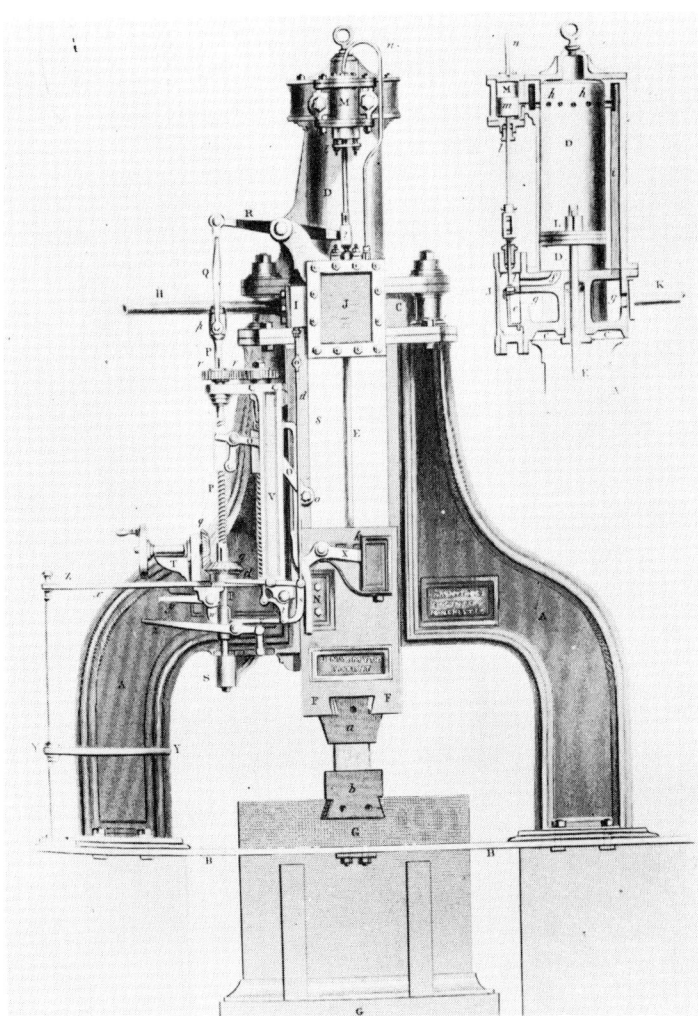

Fig. 9.3   Nasmyth's Steam hammer.

in operation the performance of these two giants was truly impressive, and their sensitivity in the hands of a skilled operator really remarkable.

The Hydraulic Press

In 1795 Joseph Bramah described an invention that was to have far-reaching applications in heavy industry. The invention in question was the hydraulic press. Essentially the equipment consists of a simple pump discharging water into the working cylinder of the press, the ratio of the dimensions of the pump piston and that of the press being such that considerable mechanical advantage results from the arrangement. This is further increased by whatever leverage is given to the operation of the force pump itself. The elements of Bramah's invention are shown in the diagram Fig. 9.4. In addition to the two non-return valves indicated there is a further manually operated valve that allows the hydraulic fluid to return to the supply tank when the ram has to be returned to its starting point.

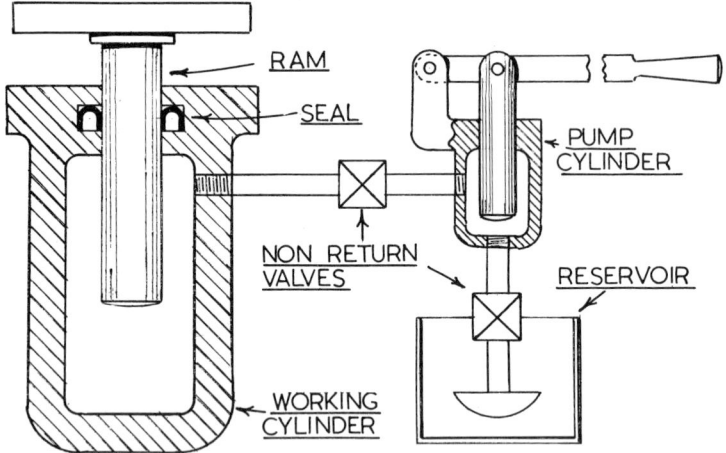

Fig. 9.4   Elements of Bramah's Hydraulic Press.

When first constructed the hydraulic press suffered from considerable leakage from the working cylinder around the ram. This condition greatly reduced the efficiency of the device and must have come near to ruining its conception. However, as so often happens, the introduction of a simple hydraulic seal put the matter to rights. Whilst Bramah himself is sometimes credited with the design of this seal, most now agree that it was Henry Maudslay, working for Bramah at the time, who found the solution to the problem.

Maudslay's seal consisted of a special-shaped ring formed from a disc of leather having its centre removed. The ring was steeped in liquid to soften it and then placed in a simple press to give it the shape depicted in the illustration, Fig. 9.5, where the seal is seen in section. When placed in a recess and surrounding the ram of the press the water pressure acts on the under side of the seal forcing the leather into close contact with both the ram and the casing of the cylinder itself. In this way any leakage is effectively sealed.

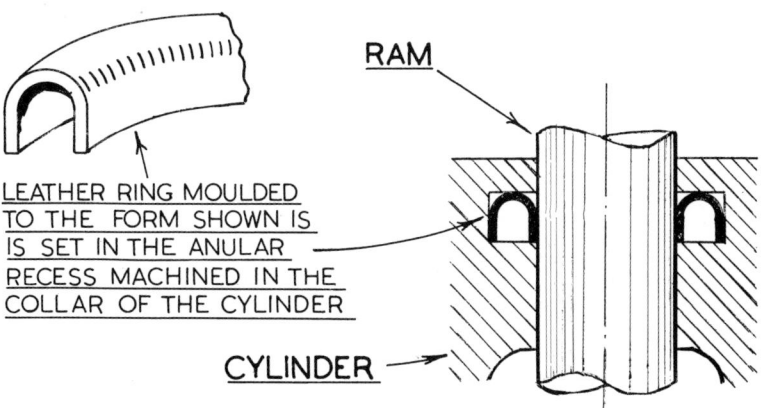

RAM

LEATHER RING MOULDED
TO THE FORM SHOWN IS
IS SET IN THE ANULAR
RECESS MACHINED IN THE
COLLAR OF THE CYLINDER

CYLINDER

Fig. 9.5   The hydraulic seal.

Since the time of Bramah's invention there have been many developments designed to increase the effectiveness of hydraulic presses and to step up their power and versatility, so that they are now used for heavy forging duties and other metal-forming purposes where presswork can be most conveniently be applied.

The first patent taken out for an industrial hydraulic press was one by Sir Charles Fox in 1847. Whilst the flexibility of the steam hammer itself was greater than that of any device that had preceded it, the ease of control of the hydraulic press far surpassed that of the steam hammer. It was natural, therefore, with the ever increasing demand for large forgings to be used in shipyards and elsewhere that advantage should be taken of the enormous power of the hydraulic press and its convenient operating characteristics.

The first press to be erected in England following on Sir Charles Fox's patent was one of 500 tons capacity at the works of Plant Brothers at Oldham. This machine was installed in 1866, some time

F

after Fox's patent had been secured, but there seems no information as to what developments had taken place in the meantime with the exception of the fact that the use of the hydraulic press for forging purposes followed the procedure for pressure forming initiated by one J. Haswell in 1861.

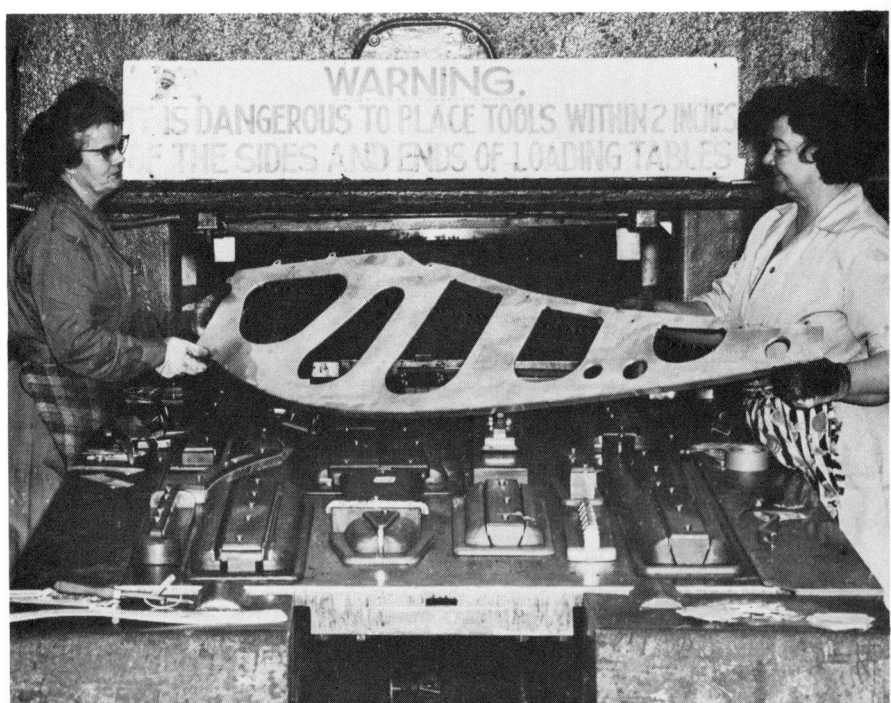

Fig. 9.6   Work carried out on the Rubber Press.

By 1887 the capacity of these presses had increased to a figure of some 4000 tons, a press of this size being erected in Sheffield for the purpose of forging open-hearth steel ingots.

An interesting development in metal-forming as opposed to forging is the employment of hydraulic power to the production of sheet metal aircraft parts. These components, for the most part, could be exercises in panel beating, but the handwork this involved is time-consuming and needs a considerable labour force to produce the parts in any quantity.

The answer to the requirement of complex sheet metal components in quantity is the rubber press. Essentially the machine is a large hydraulic press having a plain platten mounted on the ram, whilst the bolster set above it, instead of being of solid construction, comprises a mattress of rubber some 6 to 8 inches thick.

The type of component that can be finally formed by the process is well exemplified by the illustration Fig. 9.6, showing the two operatives manning the press holding a large sheet metal aircraft rib. The tools used, which may be seen in Fig. 9.7, are made from Jabroc, a hard laminated wood, and are set loosely on the loading table.

Fig. 9.7   Tools set up on the press loading table.

CHAPTER TEN

WOODWORKING MACHINERY

THE DEVELOPMENT of power-driven woodworking machines has, in general, followed the practice of adapting tools previously hand operated. Thus we have the saw, the plane and the router all becoming mechanised at one time or another. Of these the saw was the first to be power-driven.

The date when a saw mill was first in existence is somewhat in doubt, but it seems that there may have been such mills as early as the fourth century A.D., for there is an account of the erection of a saw mill in Germany on the River Roer, or Ruer, about this time with water providing the motive power.

The Pit Saw

Because it would have been a natural consequence in development to have mechanised a saw that was already being hand operated, the type used in early mills was the pit saw. In case there may be some readers unfamiliar with the tool, it should perhaps be explained that this type of saw is worked by two men with the work suspended over a pit in which one of them takes his place, while his mate stands on the top of the work, the saw being between them. The saw itself is furnished with a pair of cross handles, the sawyers using both hands to operate the tool which is alternatively pulled up and then down. See Fig. 10.1.

Saw mills were not common until the 14th or 15th century and many of them suffered the fate of so many power-driven devices at the time their introduction being opposed by hand-sawyers who saw in them a threat to their livelihood. Thus a saw mill erected by a Dutchman near London in 1663 had to be abandoned because of labour troubles. Again in 1767 the mob all but destroyed a wind-driven mill that had been set up in Limehouse.

This mill had been erected by a certain James Stansfield who had

Fig. 10.1   The Pit Saw.

gained experience, both in Holland and in Norway, in the erection and maintenance of woodworking mills.

Per contra the Scots seem to have been more receptive to mechanical invention and the benefits to be obtained from it, because it is recorded that another wind-driven mill, set up in Scotland some time before James Stansfield's ill-fated Limehouse venture, suffered no interference from labour whatever.

The type of machine saw used in the 18th century is that illustrated in Fig. 10.2, the individual saw blades being secured by means of wedges driven through lugs forming part of the frame assembly itself, and passing through holes formed in the ends of the saw blades.

The saw frame itself was usually set to make from 100 to 120 strokes per minute, a rate that hand-sawyers could not hope to maintain. The machine itself was of simple construction as may be seen from the illustration. A crankshaft with a flywheel provided the drive, the saw frame rising and falling on a pair of guide bars attached to the wooden main frame of the machine. Work was placed on a movable table which could be fed forwards in the first instance by hand and then

Fig. 10.2   Mechanical Pit Saw

automatically through the ratchet-and-pawl system seen at the near side of the main frame. The automatic feed was driven from an eccentric fixed to the end of the crankshaft, the rate of feed apparently being fixed for there does not seem to be any device for varying it.

Fig. 10.3 Mechanical Pit Saw.

The Circular Saw

There is considerable doubt as to when the circular saw was first introduced. Saws in circular form had been known since the time of Dr. Hooke, 1635–1703, but these were of small size, intended for cutting clock wheels, so they were, in effect, milling cutters.

The use of the circular saw is sometimes attributed to General Sir Sam Bentham, 1757–1831, who was responsible for the woodworking machinery used in British Naval Dockyards. He was particularly concerned with the manufacture of the wooden blocks used in rigging ships, for by the year 1793 the Admiralty were needing a total of 100,000 of these blocks annually.

It seems more likely, however, in this connection at any rate, that Marc Isambard Brunel, 1769–1849, was the actual instigator. Having approached Bentham in 1801 with a scheme for making these blocks that Bentham recognised as far superior to his own, Brunel took out a patent for sawing timber with a saw of circular form. The machine

itself likely followed the design illustrated in Fig. 10.4 which is taken
from a book published about the time, 1805, that Brunel took out his
patent.

Samuel Bentham himself was an interesting character. The only
brother of Jeremy Bentham the publicist, and of scarcely inferior
ability, he had early devoted himself to the study of naval architecture.
He had the opportunity of visiting naval establishments in the Baltic

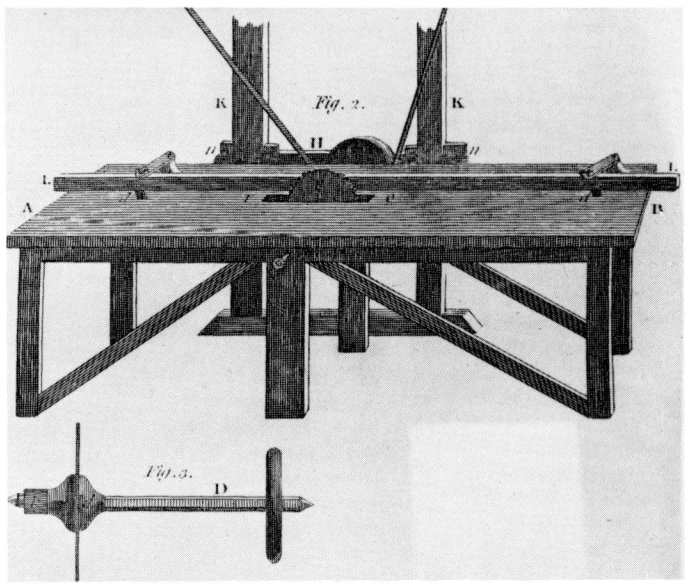

Fig. 10.4   Early Circular Saw.

and Black seas and while there, was induced to enter the service of
Empress Catherine the Second for whom he built a flotilla of gunboats
and defeated the Turkish fleet.

When the Empress died Bentham returned to England to be
employed by the Admiralty for whom he again went to Russia to
oversee the building of ships for the British Navy. Under the Admiralty, Samuel Benthan attained the rank of Inspector-General of Naval
Works a title which possibly led to him being called 'General Bentham'. It was in this capacity that he came into contact with Brunel.

The blockmaking machinery designed by Brunel and made by
Maudslay, who doubtless also had a hand in its design, was fully
operational in Portsmouth Dockyard by 1808 and the results were
truly remarkable.

With ten unskilled men performing the working of 110 skilled personnel, it was turning out 138,000 blocks annually at an estimated annual saving of £24,000.

The plant itself, one component is depicted in the illustration Fig. 10.5, was one of the first to use machine tools for mass production purposes and the design and workmanship put into it was so good that many of the individual machines remained in use for over 145 years.

Fig. 10.5   Brunel's Shaping machine.

The more interesting of them have been preserved, notably his shaping engine, which is illustrated, and his mortising machine.

Brunel made use of several types of circular sawing machines in connection with the block-making plant. Of these the pendulum saw is perhaps the most interesting. The apparatus is used for cross-cutting timber that has already been cut into plank form, and it can, of course, be employed for the production of wood blocks ready for other machining operations.

The pendulum saw consists of a frame work suspended from the ceiling having, at its upper end, a pulley system whereby the saw spindle at the lower end of the frame is driven from the main lineshafting. The arrangement is depicted in the illustration Fig. 10.6.

Until electrical power became freely available woodworking

Fig. 10.6 The Pendulum Saw.

Fig. 10.7 A wood saw bench

machinery had to be driven from the factory lineshafting or, if the saw mills was out in the country as many were, by means of a steam engine connected directly to the saw bench by a flat leather belt. Nowadays the steam engine has given way to internal combustion in one form or another, so it is not uncommon to find logging saws set up in the woods and driven by belting from the power take-off of an agricultural tractor.

The saws used were usually of large diameter and the bench itself was provided with an automatic work feed, the weight of the log for the most part itself keeping the work in place.

Fig. 10.8   Angular ripping.

The modern saw bench is completely self-contained, its electric motor and driving belts being contained within the main frame of the machine itself. This makes for much versatility because the belting is hidden and out of the way of the operator.

The saw bench illustrated, a product of Wadkin, Leicester, and seen in Fig. 10.7, has a tilting arbor enabling angular ripping and compound angular cross cutting to be carried out. The first of these activities is depicted in Fig. 10.8. In order to permit the saw to be tilted both the arbor bearings and the driving motor are mounted on a sub frame that may be set and locked at any desired angle up to a maximum of 45 degrees.

Fig. 10.9   Moulding in the circular saw.

Fig. 10.10   Trenching in the circular saw.

The versatility of this saw bench is well illustrated by its capability for moulding and for trenching as depicted in Fig. 10.9 and Fig. 10.10. The moulding operation requires the saw to be replaced by a block carrying cutters that will machine work up to a maximum of $1\frac{1}{4}$in. wide.

Trenching employs what is known as a dado saw enabling grooving to a maximum of $\frac{13}{16}$ to be cut.

The radial saw, seen in Fig. 10.11, has taken over, in some sort, the

Fig. 10.11   The Radial Saw.

Fig. 10.12 Cross cutting with the radial saw.

Fig. 10.13 Compound angle cutting with the radial saw.

duties of the pendulum saw; but it is a machine capable of a far greater range of work as the illustrations will show.

In the Wadkin Radial Saw depicted, the saw itself is mounted directly upon the spindle of the driving motor. This, in turn is carried in gimbals, the gimbal frame itself, or the yoke as the makers call it, being capable of turning through 360 degrees. In addition, the motor inside the yoke may be tilted up to a maximum of 90 degrees anywhere between the horizontal and vertical, so a wide range of compounding here is possible. Add to this that the arm, on which the saw unit is moved by the operator, may be set and locked up to 45 degrees either side of the cross-centre line of the work table, then the versatility of such a machine will be appreciated.

For straight cross-cutting the saw frame is set up as shown in Fig. 10.12, whilst the capability for compound angle cutting is well illustrated by Fig. 10.13.

In addition to the sawing operations described the radial saw illustrated can be used for moulding work.

The Mortising Machine

The purpose of the mortising machine is to cut the seating for tenons used in the assembly of wooden components such as window frames, a typical example being depicted in Fig. 10.14.

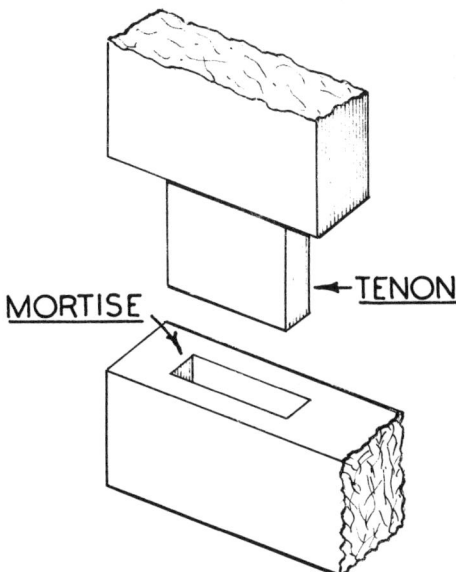

MORTISE ←TENON

Fig. 10.14   The Tenon and Mortise.

Early mortising machines were hand operated and were of the basic form shown in the illustration, Fig. 10.15, from which the method of working will be evident. The work is held in a substantial vice mounted on slides that enable the work itself to be set directly under the chisel that will cut the mortise. The chisel is mounted in a holder that can be turned through 180 degrees, so enabling the operator to cut all sides of the mortise squarely.

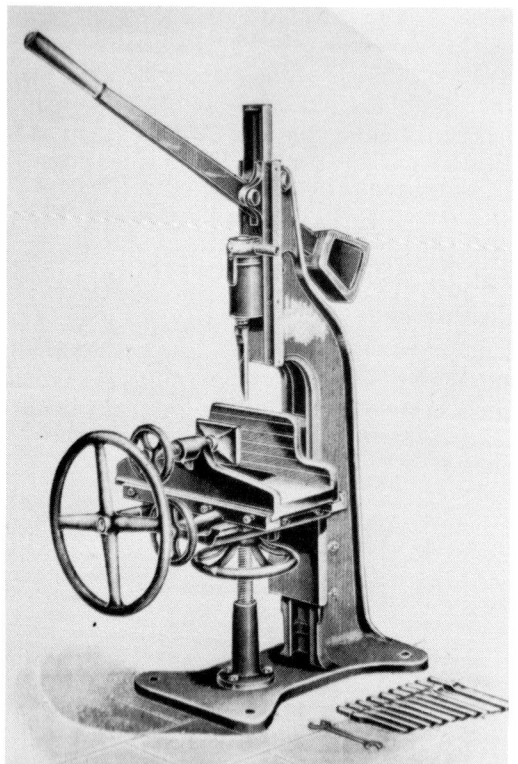

Fig. 10.15   A type of hand mortising machine.

One disadvantage of the machine is that the chisel is operating 'blind', the wood core of the mortice having to be removed by the chisel alone. For this reason a drilling attachment was added so that the bulk of the unwanted material could be removed by the augur seen mounted in the second machine illustrated. The chisel itself was then left to cut down the sides and ends of the mortise, a relatively simple and quick process that greatly increased the speed of the whole operation.

Brunel's Mortising Machine

It was left to Marc Isambard Brunel to design and make, with the aid of Maudslay, the first self-acting mortising machine. This formed one of the components of the famous Portsmouth block making plant and is illustrated in Fig. 10.16. The machine was used to cut the seatings for the sheaves over which were passed the ropes that make up the rigging of a sailing ship. Two chisels were employed each making some 400 strokes a minute. By this means either two single blocks or one double

Fig. 10.16 Brunel's mortising machine.

block could be held in the vice and machined at one setting. The vice was mounted on a travelling carriage, leadscrew-operated by a ratchet-and-pawl mechanism and actuated by an eccentric on the mainshaft. The chisels did not cut 'blind', for Brunel arranged for a drilling operation on the blocks to be carried out so that the chisels had a clear passage through the work at the commencement of the mortising operation. The fixed rate of feed advanced the work $\frac{1}{24}$ in. for each chisel stroke and the travel of the carriage was stopped automatically at the conclusion of the work cycle.

Thus the operator had only to load the vice and move the clutch

Fig. 10.17   Brunel's mortising machine.

lever to set the machine in motion; after this the remainder of the work was fully automatic.

The chisels themselves were carried in a frame travelling in guides and crank-operated from the mainshaft of the mortising machine, the working stroke being fixed.

It is said that Brunel's Mortising Machine gave rise to the idea of the slotter, already referred to in Chapter 7; this may well be so for all the basic mechanical movements needed in its metal working counterpart are present in the woodworking machine Brunel designed.

Drilling Machines for Woodworking

In the woodworking field the amount of power needed to drill holes, even quite large ones, is comparatively small. Therefore, the drilling heads of the machines used are usually of a light type capable of a high spindle speed. Typical of these speeds, as applied to the light drilling machines made by the South Bend Lathe works of America, for example, are a range of four speeds from 710 to 4,470 revolutions per minute. These machines are, of course, designed for drilling metal as well as wood, the speed range being chosen to embrace all possible materials efficiently.

Historically, the drilling machine for woodworking has followed the lines of the other devices for drilling purposes described in Chapter 8. But it seems possible that it was Brunel who appreciated the need for high spindle speeds when working wood. He had included two machines that qualify for the title of drilling machine though called by other names, amongst the block making equipment installed in Portsmouth Dockyard. They were designed for special purposes and no way resemble the normal drilling machine now used.

An example of a light drilling machine is seen in Fig. 10.18. Equipment of this class is particularly suitable for woodworking for not only can it be used as a drilling machine but also, when a hollow chisel attachment is fitted, the equipment may be employed for mortising.

The mortising unit is illustrated in Fig. 10.19 and will be seen to consist of a stirrup carrying the hollow chisel fixed to the quill of the drilling machine and a fence with clamps bolted to the work table. A true-running augur bit, a close fit in the chisel, is set in the chuck and this, when run at a high speed, removes the bulk of the wood in the mortise, leaving only the corners to be cleaned out by the chisel. The fence is readily adjusted to bring the chisel into alignment with the work, allowing the operator to maintain a true line when cutting the mortise.

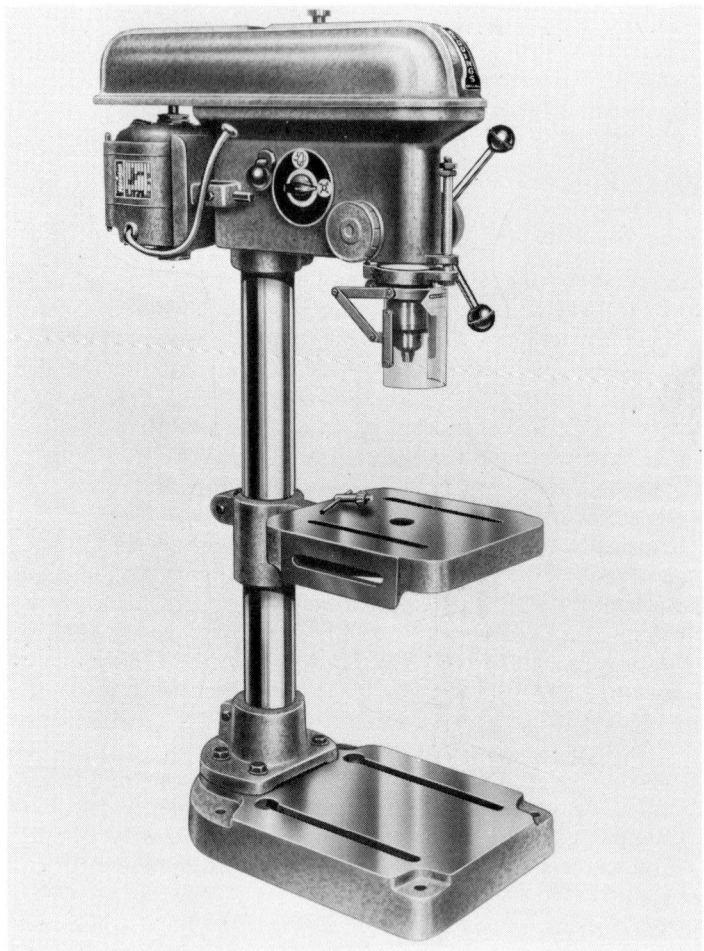

Fig. 10.18 The Pacera drilling machine

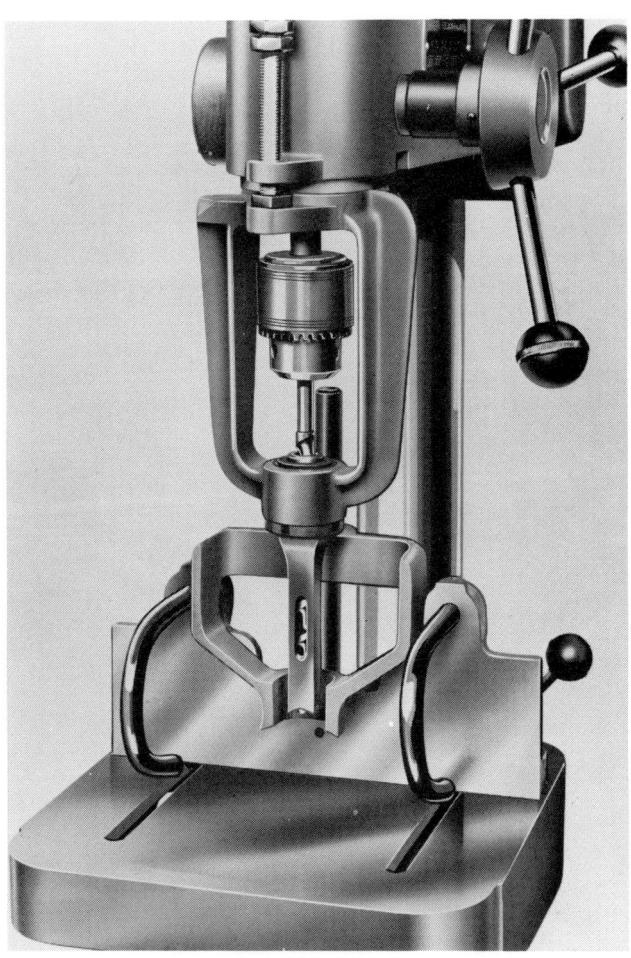

Fig. 10.19   The mortising attachment.

The Spindle Moulding Machine

In the building industry there is always a requirement for wooden mouldings. These are used for purely decorative purposes. At one time small mouldings could be produced by hand using a plane fitted with the appropriately shaped blade. Naturally, this was a laborious process producing neither the quantity nor the size of the mouldings required.

The machine tool used for producing them consists of a box-form casting carrying a work table through which projects the spindle carrying the cutter block. This spindle revolves at a high speed and the machine itself is one of the more dangerous of the woodworking units because of the difficulty of guarding the cutters adequately. Early machines carried the spindle in plain bearings but modern equipment employs ball races as befits the high spindle speed required.

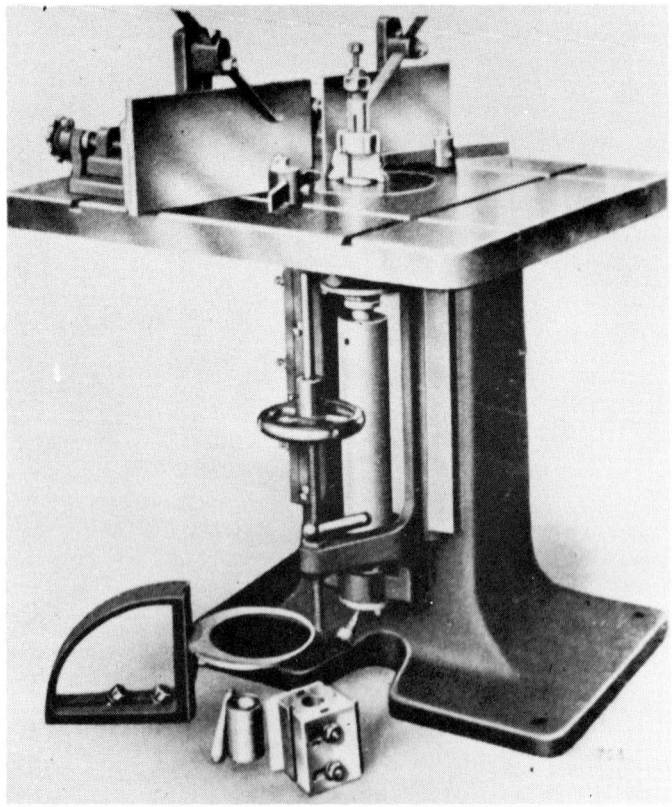

Fig. 10.20   A typical spindle moulding machine.

A vertical fence, divided to allow the cutters to project, is fitted to secure a controlled and even depth of cut. A typical machine made about 1930 is illustrated in Fig. 10.20.

The cutters are made from flat tool steel, filed to shape and then sharpened to a keen edge. They are secured to the cutter block by bolts passing through a slot or slots machined in the cutter itself. A typical cutter is depicted in the illustration Fig. 10.21.

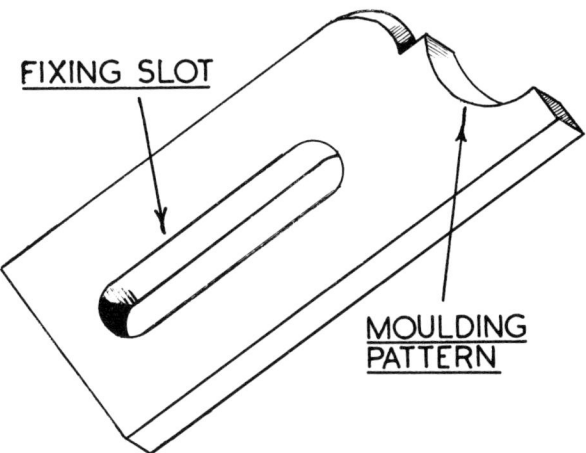

FIXING SLOT

MOULDING
PATTERN

Fig. 10.21   A spindle moulding machine cutter.

The Band Saw

From the point of view of the operator the circular saw is limited in its scope in that it can only cut in straight lines. When it becomes necessary to saw curves then the band saw is the tool to be used. The saw fitted is made from a length of narrow tool steel made endless by a brazing operation. The depth of the band varies and so does the thickness, both these dimensions depending on the size of the actual machine to which the saw is fitted. The band runs on rubber-tyred wheels one of which has means of movement vertically in order to adjust saw tension. It will be appreciated that the saw itself must be able to withstand any twisting motion that may be imparted by the actual operation of sawing. The restraint required is imposed by side rollers set above and below the table, the upper rollers being mounted on a standard that can be adjusted to bring them as close to the work as possible.

154

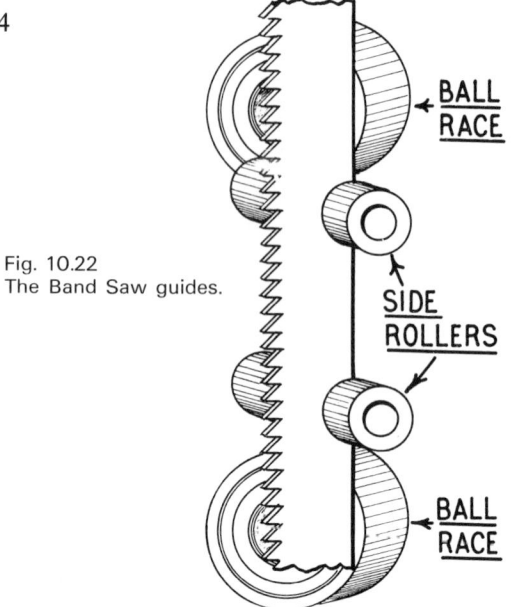

Fig. 10.22
The Band Saw guides.

BALL RACE

SIDE ROLLERS

BALL RACE

Fig. 10.23   The Band Saw.

The arrangement is depicted diagramatically in Fig. 10.22. In addition, in order to absorb the end thrust applied to the saw when in work, a modern band sawing machine is fitted with the two ball races also shown in the illustration.

The machine illustrated in Fig. 10.24 was made about 1888 and is intended for hand operation. The work table can be tilted for cutting on the bevel and the saw itself is provided with restraints above and below the table to prevent displacement during working.

Many medium-size band sawing machines today are of fabricated

Fig. 10.24   A hand-driven Band Saw of 1888.

Fig. 10.25 Machine Band Saw about 1880.

Fig. 10.26 The Wadkin 20 inch Band Saw.

construction, that is they have main frames made from sheet steel, reinforced and welded, instead of iron castings used heretofore.

The machine illustrated in Fig. 10.26 is an example of fabrication as applied to the construction of machine tools. It is also interesting in that the driving motor, housed in the base of the main frame, has the lower saw wheel mounted and keyed directly to its rotor, thus avoiding the use of any belting in the driving mechanism. The saw wheels themselves are made from light alloy so as to reduce inertia, and a foot-operated brake is provided to quickly bring the saw to rest between successive operations. As might be expected much care is taken to exclude sawdust from all working parts. This, for the most part is achieved by totally enclosing the 2h.p. driving motor and by fitting sealed ball races wherever bearings are required.

The Lathe

For woodworking purposes the lathe seems certainly the first machine tool to have been produced. It followed the lines of similar tools used for metal turning, the same lathe being used for both classes of material. In woodworking, for the most part, the turning tool has usually been guided by hand though a simple compound slide rest can

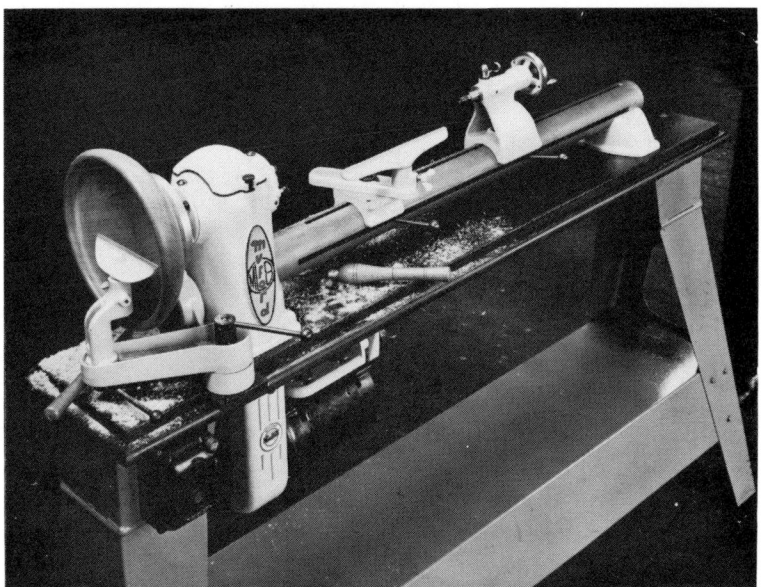

Fig. 10.27 The Myford ML 8 woodworking lathe.

often be employed as an aid in achieving parallelism and the maintenance of uniformity of size when a number of small parts are involved.

The modern wood turning lathe is exemplified by the Myford ML8 illustrated in Fig. 10.27.

This is a round-bed machine with a ball-bearing headstock capable of the high spindle speed needed for efficient wood turning.

The bed itself is of sufficient length to permit details such as table legs to be mounted for turning, while the mandrel is extended at its outer end so that a faceplate can be mounted when fruit bowls and the like need to be made.

The driving motor is fitted to the underside of metal stand that supports the lathe, the drive being by means of a V-rope of the type described earlier.

CHAPTER ELEVEN

MODERN MACHINE TOOL TECHNOLOGY

IN AN EARLIER CHAPTER the subject of the automatic lathe has been considered. It is now time to describe in outline the technology that lies behind the modern machine tool. Practically every type of production tool is involved since automation is not now confined to the lathe only.

It is the computer that has mainly been at the root of the revolution and has provided the technologist with the means of massive development in machine tool construction.

For the most part the control of the machines is either by punched tape, or by magnetic tape such as is used in music and speech recording apparatus. The first tape is similar, when correctly perforated, to the rolls once used in pianolas, music players now alas defunct. The punched tape is passed through a 'reader' consisting, in the main, of electrical contacts normally separated by the paper of which the tape is composed. The contacts are aligned with lines of holes in the tape so that an electrical connection is made once these contacts have come together through a punched hole. The electrical impulses from the 'reader' are then fed to the appropriate circuitry controlling the individual movements of the machine tool. These movements are commonly hydraulic powered.

Magnetic tape provides electrical impulses similar to those used in sound recording. These are fed into an amplifier the output from thence to relays that control the machines hydraulic system.

Basic to the operation of the modern machine tool is some method of measuring the movement of the slides supporting the work. In machine tools, from the beginning, the screw thread has been the medium used to move these slides, and it has always been a simple matter to measure the movement imparted by the screw. There has always been an element of friction in the mechanical assemblies that are employed; friction between the nut attached to the slide and the screw registering with the nut, as well as friction between the surfaces of the slides themselves, has led to conditions unacceptable in modern machine tool technology.

In order to ensure that the slides do move positively, this friction has to be reduced materially. In the case of the feed screw this reduction has been achieved by making use of a device analogous to the mechanism sometimes used with sliding doors and with which many readers will no doubt be familiar.

This mechanism is illustrated in Fig. 11.1 and will be seen to consist of a series of balls moving in a circular trackway so that they can make contact with the rail upon which the door moves, the balls recirculating as it does so. This principle can be applied to a leadscrew or feed screw.

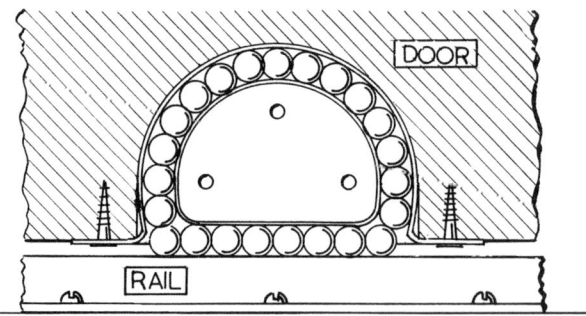

Fig. 11.1   Diagram of re-circulating ball system.

The Recirculating Ball Nut

The method used to apply the principle to machine tools is to provide their leadscrews with a recirculating ball system, to take the place of the customary bronze nut. In this way the friction, normally inseparable from the older type of assembly, is greatly reduced. The leadscrew and nut assembly is depicted diagramatically in Fig. 11.2 at 'A' whilst the modern replacement is illustrated, again diagramatically, at 'B'.

Friction in the machine slides themselves is not readily eradicated by purely mechanical methods. For this reason at least one firm has made use of pneumatics to overcome the trouble and has harnessed the air bearing to cut down the friction between the surfaces of the slide. The principle or basis of the air bearing is simple enough consisting as it does of supplying air under pressure to the joint face of the bearing assembly, for example the space between the surface of a shaft and the bush in which it normally runs; in this way the two elements of the bearing assembly 'float' in air acting as a lubricant of low viscosity imposing little or no drag on the moving part.

When the principle is applied to the slides of a machine tool the work table lifts and can be moved with little effort even if heavily loaded with the components to be machined.

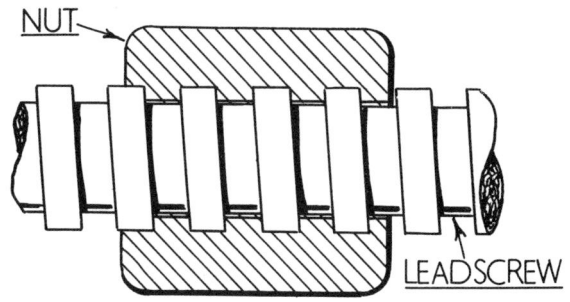

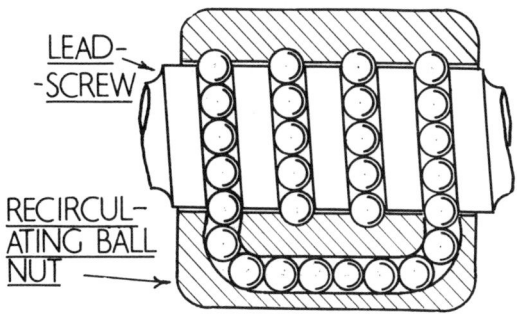

Fig. 11.2 Comparison between conventional leadscrew nut and re-circulating ball nut.

Once the work table has been moved the desired amount, the air supply can be cut off, the weight of the work and the table itself serving to help lock the slides.

Purely mechanical methods to reduce friction in the machine slides can, of course, be employed. The favourite probably is the roller track. This has already been illustrated, typically, in connection with the methods used to secure friction-low lead and feed screws, though, here, of course, it is the ball and not the roller that is employed.

In addition to the reduction of friction in the slides, it is necessary to eliminate backlash in all gearing used to drive the feed screws controlling them; otherwise accurate positioning of the slides would be impossible.

The classical method for removing backlash from a gear train, that is the elimination of play between the teeth of the individual gears, is to make one or more of the gear wheels in two parts that are spring loaded as depicted in the illustration Fig. 11.3.

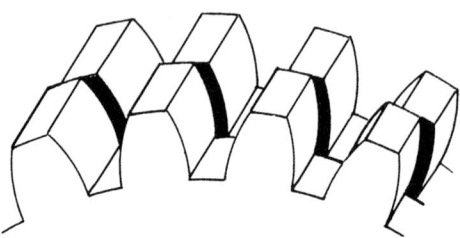

Fig. 11.3   Conventional method of eliminating back-lash in gears.

The effect is to stagger the teeth minutely in the two halves of the wheel, thus taking up any play between this wheel and the one with which it meshes. Spring-loading is commonly arranged by suitably placed tension springs or, in small sizes, by horseshoe-shaped springs similar to circlips.

What are the industrial advantages of automation? If we are condidering the type of machine that can both drill and mill as well as tap any threads required, then complete parts can be made at one setting, without recourse to a number of different machines involving a loss of time in the transfer of the parts from one machine to another.

Once a tape has been prepared, and proved on the first component machined, there is a high degree of dimensional repeatability. Lengthy inspection procedure is then avoided. As a result, when required, small batches and even single parts can be made in the minimum of time without loss of profit.

The operation of a full-automated machining centre involves the efficient use of the individual cutters and tools needed.

A numerically controlled machine of the type we have in mind is provided with the facility for automatic tool changing. These tools are pre-set by means of special equipment set up in a booth set apart for the purposes. This work does not involve a large staff for, once set, tool life is greater than with conventional machines. Provided, therefore, a sufficient supply of replacement tooling is available, and this is necessary to ensure a rapid exchange should any of the tooling break down, the skilled man in charge of the tool setting is unlikely to be overworked.

Tool changing in a representative machine takes five seconds of time only. It is carried out by a rotating arm that grips two tools simultaneously; these are the tools that need to be changed and the cutter or tool that is to replace it.

With practically every type of machine tool capable of automatic operation it is manifest that a single chapter would be insufficient space in which to describe the many machines in detail. Instead a single very versatile machine will suffice perhaps to demonstrate the variety of tooling that can now be brought to bear on the workpiece being machined.

The Milwaukee-Matic Series Ea

This machine, a product of Kearney and Trecker of Brighton, England, was designed to allow small batches of relatively complex parts to be completely machined. The operations that can be performed by it are drilling, milling, tapping, reaming and boring, and all of these are capable of performance at a single set-up.

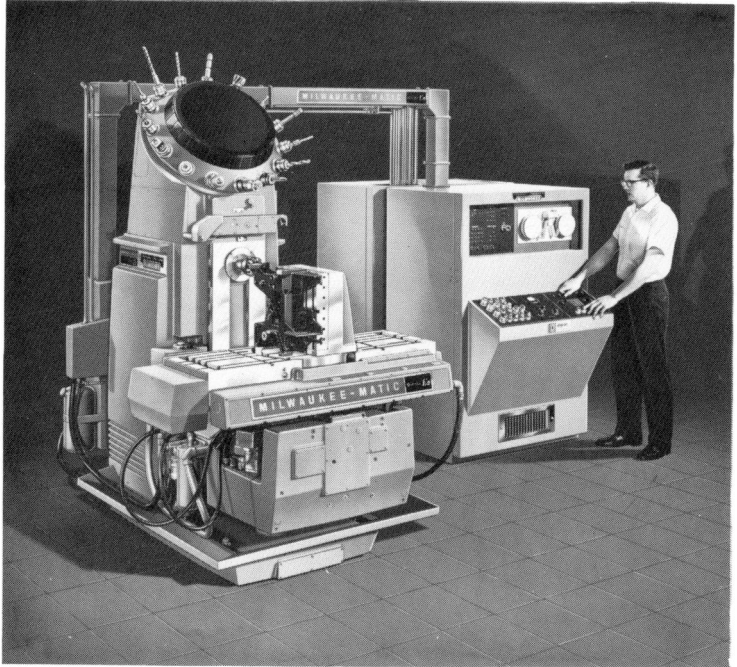

Fig. 11.4   The Milwaukee-Matic Machining Centre.

G

The machine comprises a main frame housing the essential spindle and feed screw driving arrangements and supporting the work table. In addition, as will be observed, an extension of the main frame carries the tool magazine together with the mechanism by means of which tool changing is carried out.

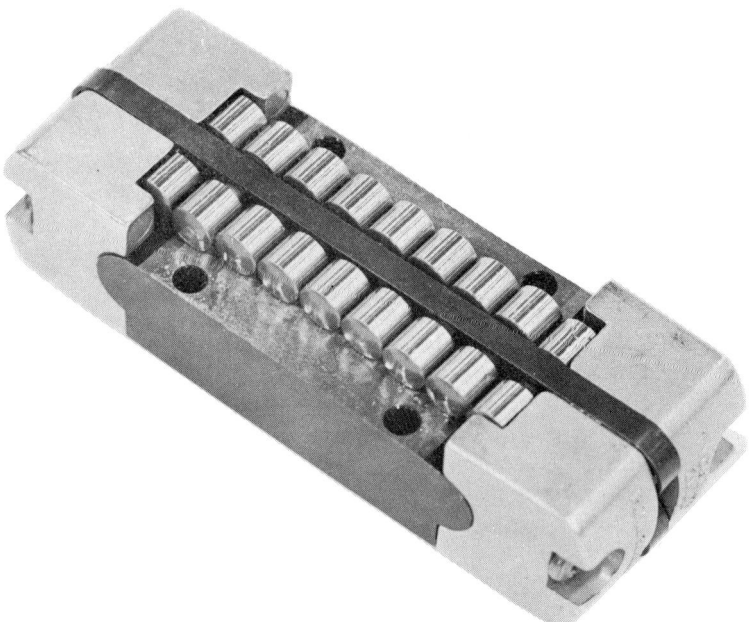

Fig. 11.5   The re-circulating roller units for the machine slide.

The complete equipment is illustrated in Fig. 11.4. Here the machine tool itself is seen in the forefront of the picture, whilst the control console and the power supplies cabinet are on its righthand side. The console itself carries the manual controls used for the primary duty of setting the machine in motion and aligning the first tool in the magazine with the workpiece mounted on the table. In addition a read-out panel is mounted above the manual controls so that the co-ordinate settings can be checked before control is passed to the tape seen to the right of the read-out panel.

The roller track system fitted to the machine slides, and already referred to earlier, consists of a number of units of the type illustrated in Fig. 11.5.

Fig. 11.6   The re-circulating ball nut and leadscrew.

These are attached to the moving elements of the slide, the re-circulating rollers reducing the slide friction to a negligible and acceptable figure.

The re-circulating ball leadscrew fitted to the machine is illustrated in Fig. 11.6.

CHAPTER TWELVE

MACHINE TOOLS FOR AMATEUR USE

THE DIVIDING LINE between the amateur and the professional, so far as machine tools are concerned, is that, for the most part the professional has access to a number of different machines, whilst the amateur, in many instances, must make do with a single unit, or perhaps two if we include the drilling machine.

The Lathe

For the amateur the lathe has always been a 'do-all', equipped to perform a number of operations in addition to turning. Even from fairly early times this has been the case. To some extent this was also once true of the professional using the smaller sizes of lathe. For example the early clock and instrument makers developed attachments for wheel cutting that had no counterpart in heavier industry.

Again the professional ornamental turners had to devise mechanisms of fairly complex form in order to carry out their projects. All this had a profound effect upon some of the machines designed specifically for amateur use, particularly in the ornamental turning field where there was, and of course still is, a well-informed and skilled body of amateur workers.

At the turn of this century there was a number of lathes designed specifically for use by workers at home, far more, indeed, than there are now.

Among the early amateurs must be numbered John Smeaton, 1724–1792, builder of one of the Eddystone lighthouses. His lathe, described in Chapter 1, illustrates the type of machine in use at the time. Of Smeaton himself James Watt had a high regard both for his practical as well as his theoretical knowledge saying 'he has made engineers of us all'.

A simple wood-turning lathe made about 1810 is illustrated in Fig.

166

12.1. Whilst this illustration is somewhat diagrammatic it does show what was then current practice. The lathe bed was of iron construction and so were all the main elements of the tool. Simple screw-cutting could be performed, the mandrel being able to slide under the control of the chasing screw seen at the outer end of the headstock.

The stand consisted of a pair of iron castings shaped somewhat like a letter 'A', the cross-member of it serving as the support for the shaft of the driving wheel whilst the feet at the back of the stand carry centres on which the treadle gear itself is mounted.

The machine illustrated is almost identical with one on which the author 'cut his teeth' as a boy. In addition to equipment seen in the illustration this lathe had a compound slide rest and was fitted with a

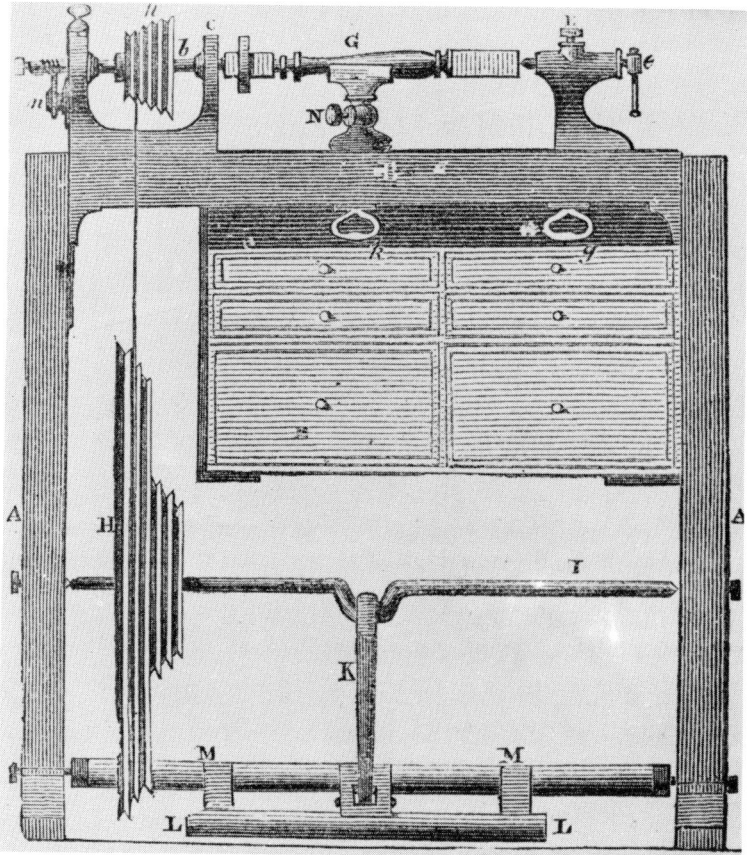

Fig. 12.1   Simple wood turning lathe.

self-centring chuck. Its owner, an uncle of undoubted mechanical proficiency, had made a number of attachments for it the better to assist him in the building of a large pipe organ, a work of herculean proportions that was some thirty years in reaching a successful conclusion.

The mandrel was solid and was hardened. It did not have the chasing equipment seen in the illustration but was provided with the bearing arrangement commonly used at the time as depicted in Fig. 12.2. The

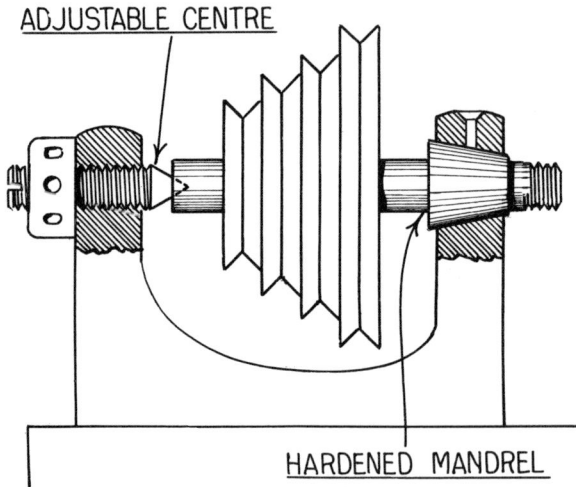

Fig. 12.2   Bearing arrangement commonly used.

mandrel itself was coned on its inner end whilst the outer end was supported on an adjustable centre enabling any slackness in the bearing assembly to be taken up. Despite many years of considerable use, memory does not suggest that any adjustment was ever needed.

Moving on from the simple amateur lathe described, the next development was the introduction of the overhead shafting seen attached to the lathe depicted in Fig. 12.3.

The basic lathe is similar to that depicted in Fig. 12.1 having, in addition, a compound slide rest capable of holding tools for turning duties or, as seen in the illustration, a milling spindle for use when ornamental work is undertaken.

This spindle is driven from the overhead shafting by means of a round leather belt provided with a tensioning device carried on the cross-member of the framework supporting the shafting. Additional pulleys are mounted on the shafting so that amongst other purposes the lathe mandrel can be driven in conjunction with the milling spindle.

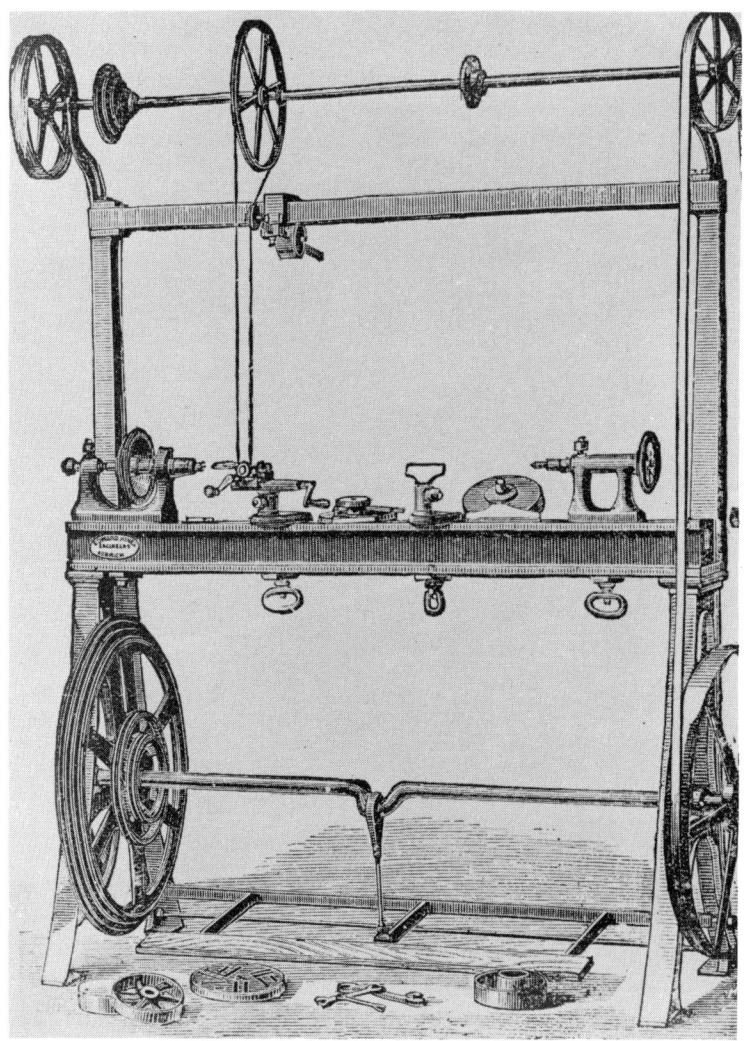

Fig. 12.3   Lathe with overhead shafting.

This facility was essential for any of the requirements of ornamental turning.

Lathes of this type were provided with fairly comprehensive equipment, some of which can be seen in the illustration. Ornamental lathes could, and often did have, much complicated and expensive equipment such as geometric chucks and rose engines; by means of these devices the turner could extend the complexity of his ornamentation.

An essential device in lathes of this class was some means of indexing the mandrel so that it could be rotated in equal steps. In its

Fig 12.4   Holzapffel Lathe 1815.

simplest form this device comprised a plate attached to the mandrel and having a row or rows of equally spaced holes that could be engaged by a detent or pin to hold the mandrel securely during the machining process. The operative selected a row of holes divisible by the number of steps required then, marking off the spacing required on the division plate, placed the detent successively in the marked holes.

The making of ornamental lathes was, for the most part, in the hands of a few specialists amongst whom was Edward Hines of Norwich one of whose products is illustrated.

Ornamental lathes have been known since the late 17th century, developing into complex and highly finished mechanisms by the middle of the 18th century. They have always been of great interest to the amateur and have been used by many influential and rich devotees to the craft. An example of a lathe made by Holzapffel in 1815 is illustrated in Fig. 12.4. Holzapffel has been rightly regarded as the high priest of ornamental turning and mechnical manipulation in general; his monumental work on the subject being much sought after by students of workshop technique.

The compound slide rest illustrated in Fig. 12.5 is of interest as it exemplifies the type often used by the ornamental turner. The arrangement comprises a basic compound rest as used for normal turning combined with a quick-acting slide to carry the cutting tool used in connection with geometric chucks and other devices.

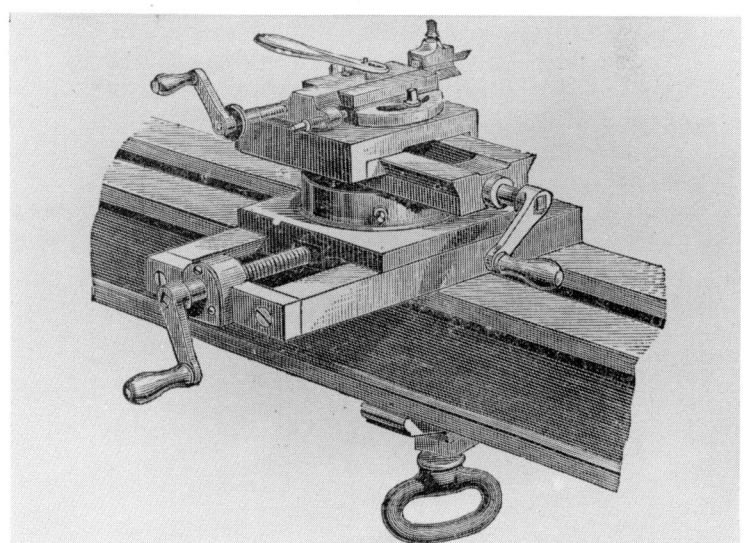

Fig. 12.5 Compound slide rest for the ornamental lathe.

By 1815 the ornamental lathe was in an advanced state. Holzapfel's equipment illustrated in Fig. 12.4 demonstrates the complexity of the attachments and the variety of the devices that were needed.

Screw-cutting Lathes

Amongst the names associated with the production of equipment suitable for use by the amateur the firm of Henry Milnes deserves to be remembered, and it is one of their early screw-cutting lathes that forms the subject of the illustration, Fig. 12.6. This tool has, in its simplest

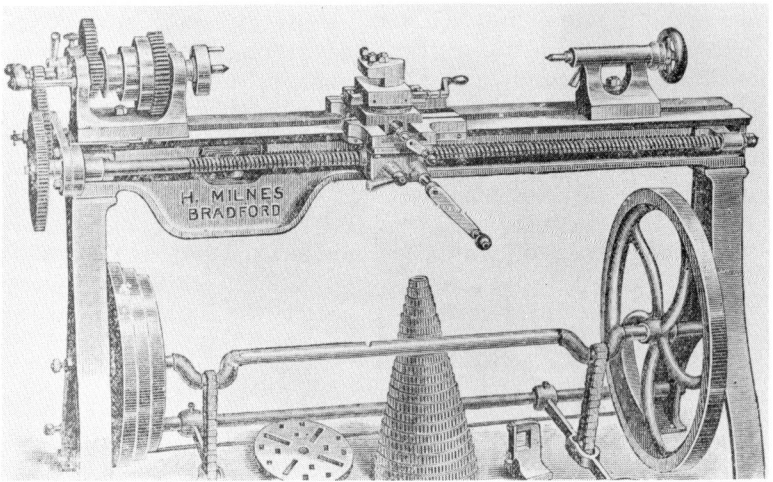

Fig. 12.6   Early screwcutting lathe by Henry Milnes.

form, all the essentials in a lathe designed for the carrying out of all normal turning operations. The machine has back gear, is provided with a full set of change wheels for screw-cutting and has the usual equipment supplied as standard for turning between the centres of the lathe. In addition, the bed is provided with a gap piece which, when removed, allows out-size work to be mounted on the faceplate. This arrangement, whilst convenient, had its penalties. The gap piece, even when accurately made and fitted, was always difficult to replace and its location, close against the headstock, contributed little to the accuracy of turning operations.

The lathe has always been a multi-purpose machine for the amateur, leading to the design and production of specialised equipment to use with it. An interesting example is that illustrated in Fig. 12.7. This is a milling slide, made by Edward Hines of Norwich, suitable for attach-

Fig. 12.7 Edward Hines milling slide.

ment to the simple, uncluttered lathe beds that were the rule at the latter end of the 19th century. The design deserves to be resuscitated; unfortunately a lathe of substantial centre height is needed to accommodate the device or its capacity would be much reduced. The device consists of a base casting fitted with two slides, one of them used for elevating the cross slide carrying the work table. In this way the depth of cut can be set accurately.

The Drummond Range of Lathes

At the turn of this century the range of machines available to the amateur was considerable. Among the lathes those made by Drummond of Guildford were probably as important as any. The first of these was the 4-in. round bed machine illustrated in Fig. 12.8. This lathe was specifically designed for the amateur; it had a round bed, was

Fig. 12.8   The Drummond 4-inch round bed lathe.

capable of screwcutting, but had no back-gear though one was later available from an outside manufacturer. The lathe was of substantial construction and was well engineered. It could be provided with a stand complete with treadle gear or it could be mounted on a wooden bench with a foot motor set below it. The standard of the ancillary equipment available was high and the whole conception a robust tool capable of any work within its capacity the model maker might care to undertake.

The lathe, however, suffered one defect. Owing to having a round bed, it was not possible to use the normal methods for guiding the

Fig. 12.9   The Drummond $3\frac{1}{2}$-inch centre lathe.

saddle. Instead a key set in the saddle, and engaging a long slot machined in the underside of the bed, served to maintain alignment. Any wear on the key would, of course, disturb this.

On the other hand, the Drummond $3\frac{1}{2}$-in. centre lathe, illustrated in Fig. 12.9, had normal methods of saddle guidance. This lathe was introduced early in the century, surviving with few modifications until about 1940 when at the request of the war-time British Government, manufacture was taken over by the Myford Engineering Company of Nottingham. After the War manufacture was suppressed, Myfords wishing to concentrate on a successful lathe of their own design.

The Myford ML7 Lathe

This lathe, of $3\frac{1}{2}$in. centre height, was introduced to meet the requirements of the amateur and the small professional worker. It followed the production of a smaller lathe, having less centre height. The ML7 lathe is depicted in the illustration Fig. 12.10, where the basic version of the Super Seven machine is seen. In common with modern lathe practice the machine has an in-built countershaft system complete with clutch, and can be supplied with a Norton gearbox to accelerate screw thread cutting. A very full range of equipment is available for this lathe which has found its way into many professional as well as amateur establishments.

Fig. 12.10   The Myford ML7 Super Seven lathe.

The home wood worker is also catered for in the Myford range. The ML8, already described in Chapter 10, is specifically designed to provide continuously the high mandrel speeds needed for wood turning. As may be seen from the illustration, Fig. 12.11, the plain headstock is provided with a pair of angular contact ball races capable of adjustment when needed. The mandrel itself has a nose at each end so that work may be mounted at either point if required. This facility is

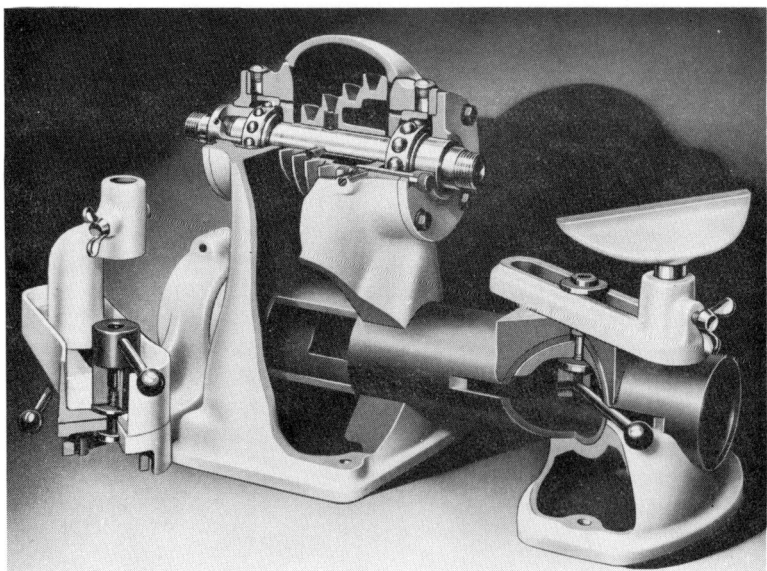

Fig. 12.11   The Myford ML 8 Woodworkers lathe.

of importance when the turning of large diameter work is undertaken; in which case the hand rest seen in its normal position, is transferred to the adjustable fitment attached to the round bed of the lathe at its outer end.

The Boxford Lathe

The great increase in interest and proficiency in the amateur field has led to the need for lathes larger than was the rule for many years. An example is the $4\frac{1}{2}$in. centre Boxford lathe illustrated in Fig. 12.12. As will be appreciated the machine has a very complete specification. The mandrel is carried in roller and ball races and there is a novel speed variation device, controlled by the wheel seen at the front of the stand

and having the resulting mandrel speed indicated on the dial seen to the left of the control wheel.

Both sliding and surfacing feeds can be made automatic, the lead-screw itself providing the drive. This it does by means of a key sliding in a keyway running the length of the leadscrew. The key engages a train of gears built into the apron of the lathe, the gear train required being capable of selection by the detent-fitted lever seen at the front of the apron. In this way the leadscrew threads themselves are relieved of all duties except their proper function of controlling the pitch of any screwcutting that needs to be carried out. The clasp nut engaging and disengaging the leadscrew is operated by the cast lever at the right of the apron.

The lathe is fitted with a built-in suds system.

Fig. 12.12   The 4½-inch Boxford lathe.

The Drilling Machine

Whilst the amateur has always been able to use his Lathe for drilling purposes if need be, the horizontal attitude of the drill entailed by this procedure does not make for comfortable working. A separate drilling machine, therefore, has always been the first addition the amateur makes to his workshop equipment.

In the past several small drilling machines have been available, some

Fig. 12.13   The ¼ in. 'Champion' drilling machine.

soundly engineered but many of very doubtful quality. Of the machines available early this century the $\frac{1}{4}$in. capacity 'Champion' drill, available at the nostalgic price of £3-15-0, was one of the best produced machines on the market.

The machine is illustrated in Fig. 12.13 where an example belonging to the author is depicted. This particular sample is fitted with mechanism to provide an extra range of spindle speed. This device was a product of the author's workshop and was not obtainable commercially.

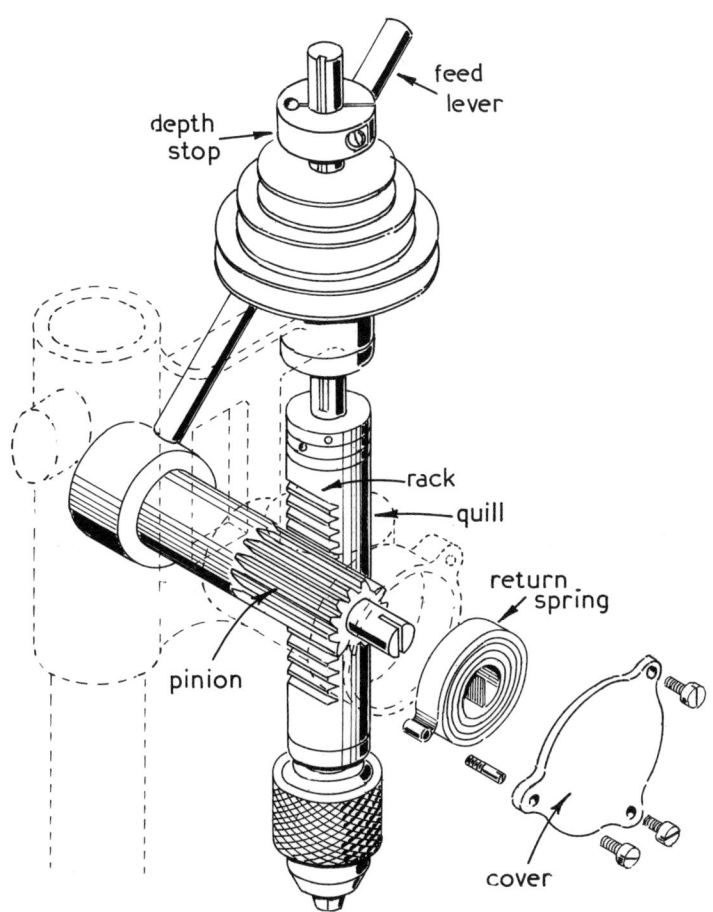

Fig. 12.14   The working parts of the 'Champion' drill.

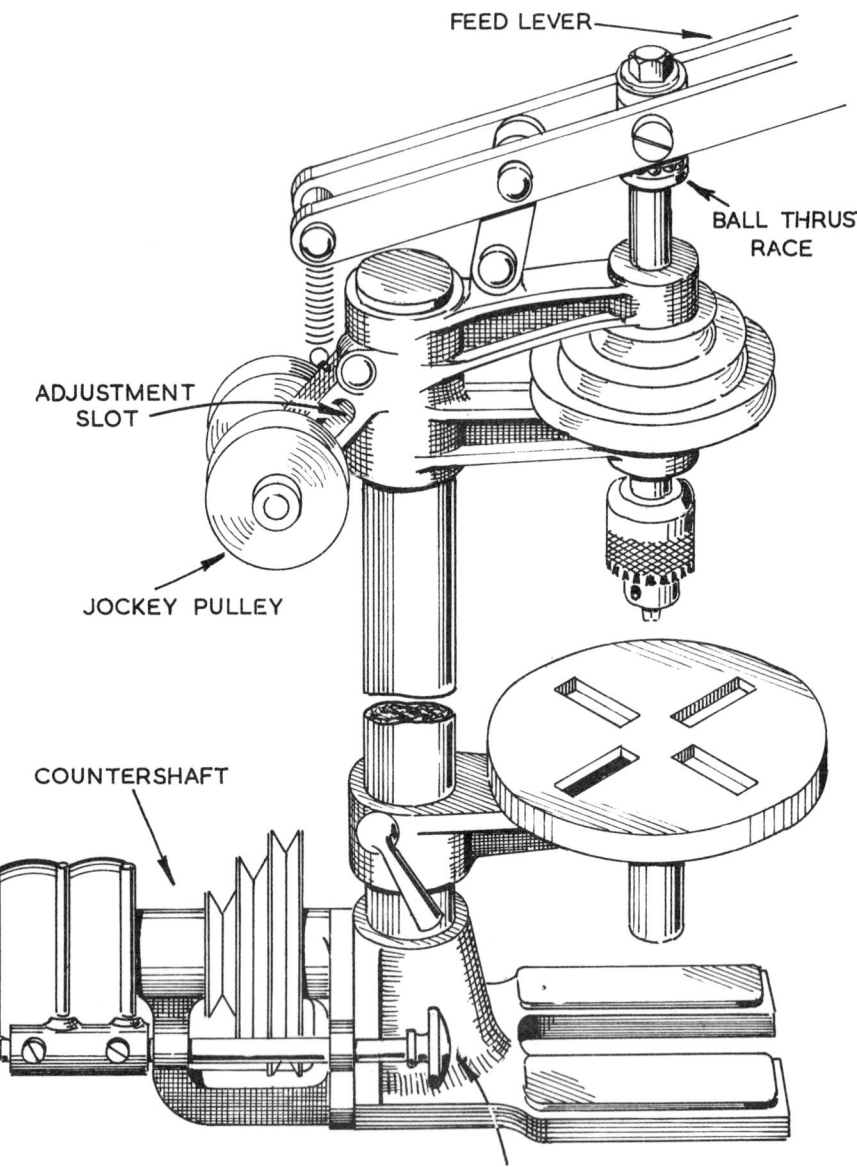

FEED LEVER

BALL THRUST RACE

ADJUSTMENT SLOT

JOCKEY PULLEY

COUNTERSHAFT

BELT STRIKING CONTROL

Fig. 12.15   The 'Model Engineer' drilling machine.

The machine itself was a quill-type drill; that is to say the drill spindle is carried in a long bearing, or hollow quill to give it the correct engineering name. As may be seen in the illustration, Fig. 12.14 showing the machine's component parts, the quill is provided with a rack which is engaged by the pinion machined on the spindle of the feed handle. An adjustable collar, fitted over this spindle, is engraved in divisions representing $\frac{1}{32}$in. of drill movement so that the operator can gauge the depth of drill penetration. In addition a stop, attached to the main spindle, allows the machine to be set to drill to a definite depth when a number of similar holes have to be formed. A spring, set in a box cast in the main frame, is used to return the spindle after drilling has been performed.

The circular work table is slotted to allow work to be secured by bolts and may be tilted to allow angular drilling to be carried out. An example of this 'Champion' drilling machine has been among the equipment in the author's workshop for nearly 40 years. Despite

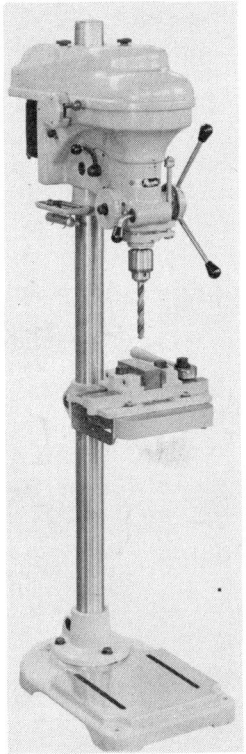

Fig. 12.16 The Pacera drilling machine.

considerable use the machine is in excellent condition, a fact that would seem to underline the desirability of its revival.

There has always been the desire of the amateur to build his own drilling machine; for this reason a number of designs, capable of production within the resources of the home workshop, have been produced. In the smaller sizes the $\frac{1}{4}$in. machine, designed by the late Edgar Westbury, is probably the best known. This machine is known as the 'Model Engineer' drilling machine and is illustrated in Fig. 12.15. Here the spindle runs directly in the head casting and the feed arrangements are, advisedly, of a simple nature. The machine, if required, can be provided with the countershaft seen in the illustration. This is a separate unit that can be machined independently and then be bolted directly to the base casting.

A somewhat larger machine of the quill type is $\frac{3}{8}$in. drill made by E. W. Cowell of Watford. This is a high-class tool provided, as will be seen, with the driving motor platform set directly behind the drilling head. A simple but effective belt-tensioning device is incorporated to ease the changing of the driving belt from one step of the pulleys to another.

For those who need a machine that will drill holes up to about 1in. diameter, the Pacera pillar drill, illustrated in Fig. 12.16, has found favour. This machine is provided with back gear similar to that fitted to many lathes. The Pacera drilling machine illustrated has been part of the equipment of the author's workshop for many years. Its driving

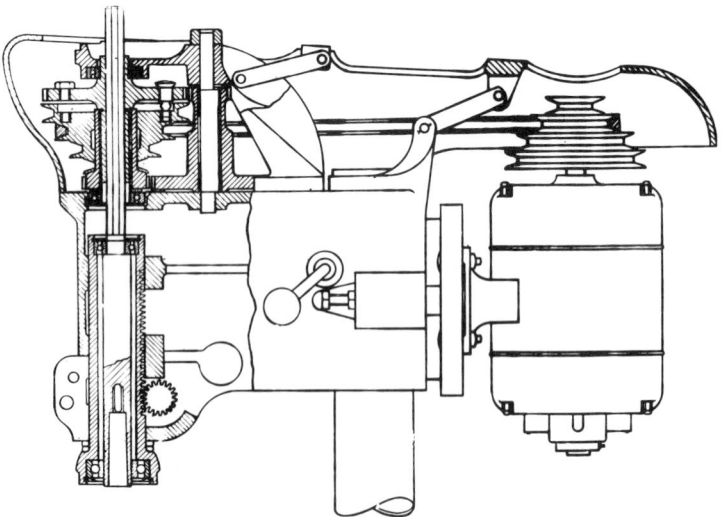

Fig. 12.17   Head of the Kerry drilling machine in section.

arrangements follow closely those of the Kerry machine depicted in the sectioned illustration, Fig. 12.17. This shows clearly the disposition of the back gear and the arrangements for the belt drive. Also shown are the quill in section and the rack feed applied to it.

Shaping Machines

In the amateur workshop, when funds allow, the shaping machine has been the third major unit to be installed. When such machines, usually hand powered, became available in quantity is somewhat uncertain. But early in this century Drummond of Guildford introduced a small hand shaper suitable for bench mounting, a cast-iron stand being obtainable if needed. This was a sturdy tool, of sound engineering design, capable of machining a surface approximately 7in. × 7in. Unfortunately, the Drummond shaper appears to have gone out of production sometime before 1940.

The shaping machine manufactured by E. W. Cowell of Watford did much to fill the gap left by the demise of the Drummond. As may be seen in the illustration, Fig. 12.18, the Cowell machine is of the travelling head type, the work table being stationary. Castings are available for this shaper, some in the final machined condition, so that the amateur worker can build the shaper for himself. Its capacity is

Fig. 12.18   The Cowell shaping machine.

much the same as was that of the Drummond, and its quality compares more than favourably with that high-class machine. Like the Drummond it is hand operated.

It is manifestly only possible to keep prices down when machines are hand powered. Nevertheless, there are those who feel that the much-increased cost of applying power to the shaping machines used by amateurs can be justified by the facilities offered.

A simple powered machine is that seen in the illustration Fig. 12.19. This tool is made by the Perfecto Engineering Company of Leicester, and is again in the travelling head class. Automatic Feed is, of course, provided and a simple method of setting the length of stroke is fitted to the driving gear itself, up and down which the crankpin can be moved

Fig. 12.19　The Perfecto power shaper.

and locked where required. This machine follows the lines of those set out by Whitworth and described in Chapter 7. There are other small shaping machines in the power-driven class, but these are really more of interest to the professional worker and follow the designs described in Chapter 7.

Milling Machines

When the amateur needs to make use of milling as a machining operation, for the most part it is the lathe he employs to carry it out. But, despite many attachments designed to assist, and these include milling slides and separately driven devices, there are many obstacles to satisfactory working.

In the past a number of small, simple, milling machines were produced; some of these are now unobtainable, whilst some others have undergone much modification leading to an increase in cost that brings them into the professional class.

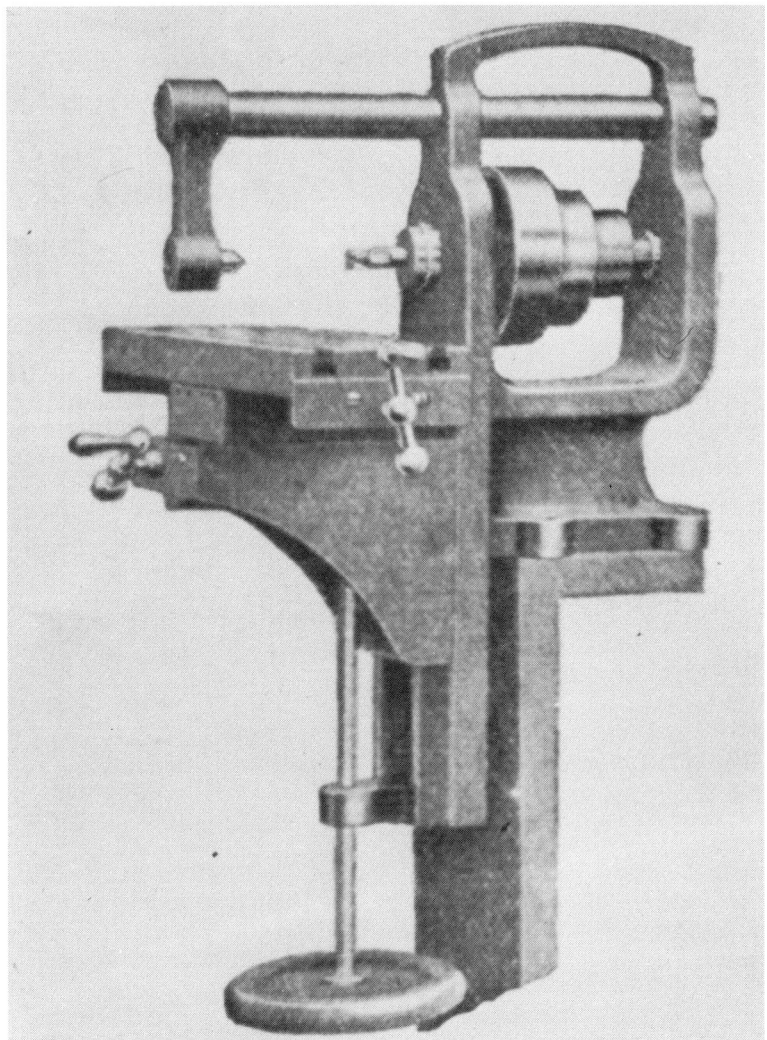

Fig. 12.20   The Pool milling machine.

Fig. 12.21   The Senior milling machine.

In the simple class the Pool milling machine illustrated in Fig. 12.20 is an obvious example. This tool has only hand feed to the work table and the spindle has no back gear. It is essentially a bench machine capable of the simple milling operations the amateur needs.

At the other end of the scale the machines made by Tom Senior have all the facilities that are required, together with the additional equipment enabling all classes of milling to be carried out. A typical machine made by the Tom Senior is illustrated in Fig. 12.21. This is the 'Junior' miller, aimed at a market comprising both the amateur and the smaller professional workshop. The machine has power feed to the work table, and can be provided with a suds system to lubricate and cool the cutter during machining operations.

Micrometer Index Dials

In common with machines made for professional purposes, tool slides and work tables are usually provided with micrometer dials enabling the operator to measure the amount of tool feed or work movement accurately. An example of these fittings is to be seen in the illustration Fig. 12.22. Here the compound milling slide made by E. W. Cowell has micrometer index dials fitted to both feed screws.

While the dials fitted to machines destined for the professional workshop, for the most part have means of adjustment so that they

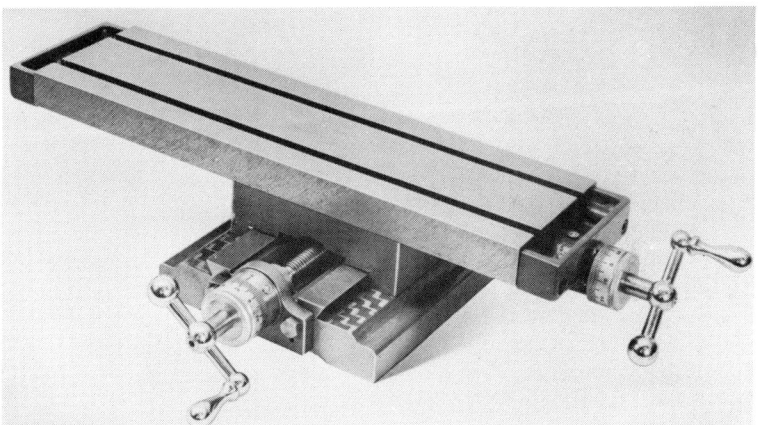

Fig. 12.22   The Cowell milling slide showing dial indexes.

may be set to zero when needed, those used with amateur lathes and the like are fixed and normally incapable of independent movement. A typical arrangement is depicted in Fig. 12.23 at A.

Here, the index dial is attached directly to the feed screw and is secured by a lock nut. The device illustrated at 'B' has been fitted by the author to several machines in his workshop. The index is mounted on a thimble that is itself secured to the feed screw. The index is therefore free to turn on the thimble but has some frictional constraint imposed by the spring-loaded device seen enlarged in the illustration. Here, a bronze plunger engages the thimble, the amount of spring pressure being adjusted by the screw holding the spring in place. If the dial needs to be locked then the screw is turned so that the spigot machined on it and that formed on the bronze plunger are brought into contact.

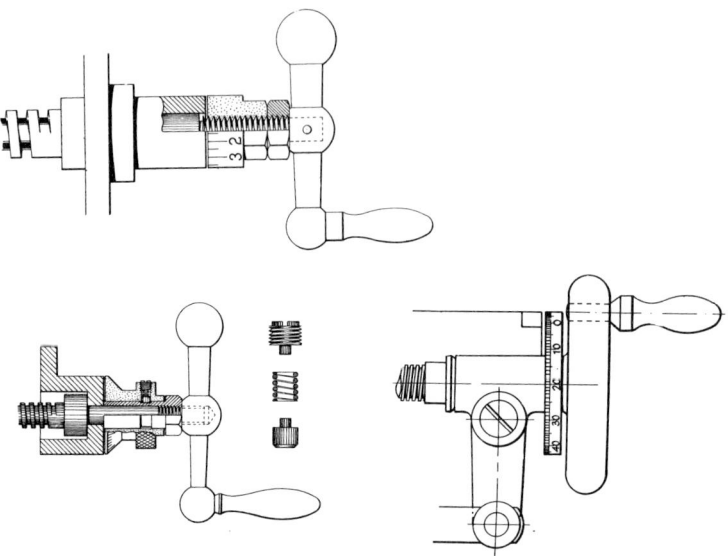

Fig. 12.23   Three forms of index fitted to amateur machines.

The index dial depicted at 'C' is one sometimes fitted to the leadscrew of amateur lathes. The index is usually non-adjustable though some amateurs have made dials that can be adjusted. There are a number of uses for the leadscrew index notably in milling and the machining shouldered work where it may be necessary to turn the component to a definite length.

CHAPTER THIRTEEN

HACKSAWING AND COLD SAWING MACHINES

THE SAW HAS ALWAYS been a normal method of cutting metal. Having been in use for centuries as a woodworking tool it was inevitable that the saw should be used in metalwork.

Handsaws were first employed, but freehand methods soon had to give place to saws guided in some simple manner in order that greater accuracy could be achieved. The equipment usually consisted of a simple frame with slide to carry a hand hacksaw, and a vice, mounted on the machine's base, to grip the material to be cut.

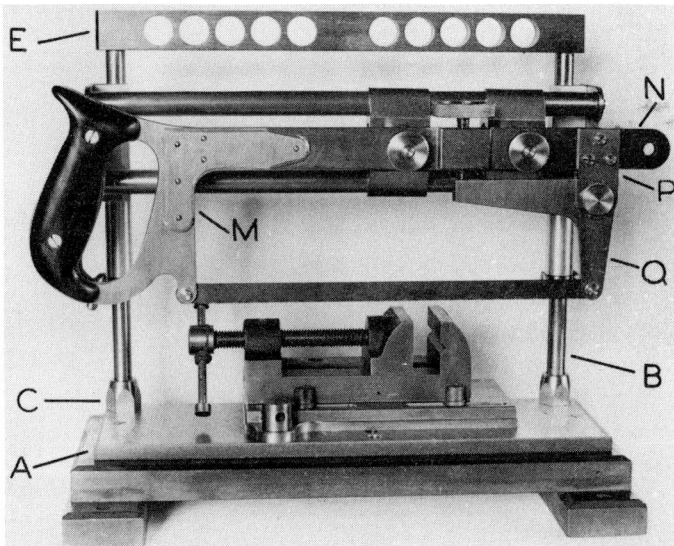

Fig. 13.1   Hand Hacksawing machine

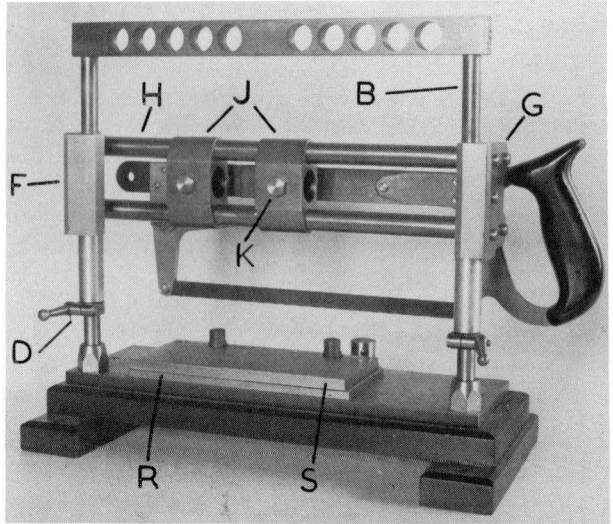

Fig. 13.2   Hand Hacksawing machine.

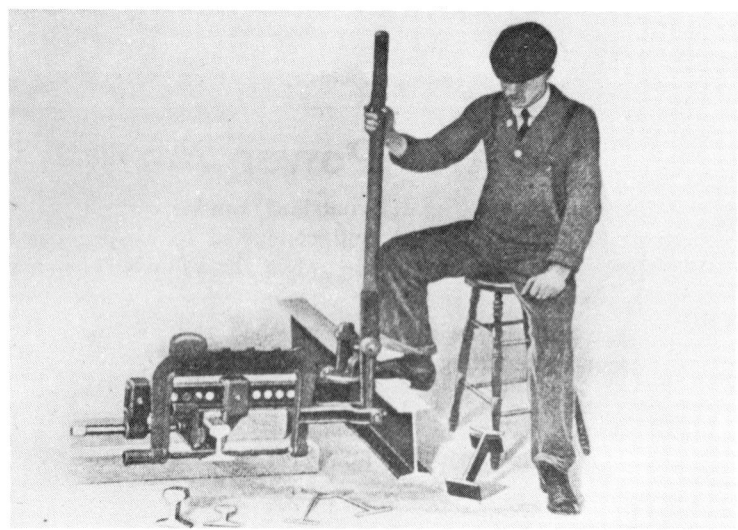

Fig. 13.3   A large handpower sawing machine for railway duty.

The illustrations, Fig. 31.1 and Fig. 13.2, depict such a device as built by an amateur constructor. It comprises a base 'A' carrying two pillars 'B' and 'C' joined together at the top by the cross member 'E' whose purpose is to retain the uprights parallel with one another. A pair of runners 'F' and 'G' are mounted on the uprights and these serve as mounts for the cross slide bars 'H' that carry the slides 'J' to which the saw frame itself is fixed.

A pair of adjustable stops 'D' are fitted to the uprights; these serve to set the depth to which the saw will cut when this is necesary.

A vice for holding work or material to be sawn is attached to the base 'A' and may be adjusted for position or angle by means of the mountings 'R' and 'S'.

Constructional industries, and in particular Railway companies, have for many years made use of large hand-driven sawing machines, usually mounted on the ground, when cutting lengths of rail or rolled steel joists. A typical example is illustrated in Fig. 13.3. This illustration appeared in a tool merchant's catalogue dated 1912 and is very representative of equipment that many may remember seeing by the side of the railway line at one time or another.

Simple Power Driven Machines

Industrial requirements in the early part of the last century led to the production of simple machines capable of cutting off the lengths of material needed for production purposes. Early hacksawing machines do not seem to have been provided with any positive means of relieving the saw blade on the back stroke. When a heavy load is applied to the teeth of the blade they tend to become blunted if no relief arrangements are made. On the other hand, lightly loaded saw teeth seem not to wear unduly. Nevertheless, the advisability of providing relief was soon appreciated and was achieved, in the first place, by a a simple cam operated mechanism lifting the saw arm on the return stroke.

Power hacksaws were, and still are, principally used for cutting off raw material. The speed of operation was based on the number of strokes per minute that a workman with a handsaw could make comfortably.

A rate of some 60–80 strokes per minute was normal as best suited to the class of blade available at the time.

In a large works the power saws are usually located in the raw material stores. This being so, the machines are often called on to produce a number of pieces of material all identical in length, a requirement that led at least one firm to produce a power hacksaw that would take a length of material and cut it up automatically. The machine would accept the metal bar, move it the required amount between cuts, and ring a bell for assistance when the work had been completed or if a saw blade had broken.

Whilst the use of floor-mounted sawing machines continued for many years it is only recently that bench machines have become at all common. They owe their popularity to their usefulness on the fitting bench, and in the small workshop, where their versatility and capability of carrying out work of a varying nature is much appreciated by many experienced workers.

Fig. 13.4   Bench hacksaw.

It seems that bench hacksawing machines may have owed their origin to two amateurs who built their own equipment over 25 years ago. A test machine was built first, and this is depicted in Fig. 13.4. The final design is illustrated in Fig. 13.5. Drawings for the machine were published leading to its construction by other amateur workers. Perhaps as a result, the advantages of a practical bench machine became apparent to some commercial undertakings who arranged to supply either complete machines or partly finished components enabling the eventual user to build the machine for himself.

The Cowell machine depicted in Fig. 13.7, is made by E. W. Cowell of Watford and is an example of a hacksaw that may be obtained in kit form whilst the unit seen in Fig. 13.8, is made by Pacera of Slough who only supply complete machines. Whilst the amateur-designed hacksaw provided no relief for the saw blade both commercially made machines do so.

Fig. 13.5   The amateur-built bench hacksawing machine.

Both professionally made machines have oil dashpots. These are devices that allow the user of a hacksaw to lower the saw blade on to the work gently without fear of snatch. A dashpot system consists essentially of a cylinder, in which a piston attached to the slide of the saw frame is free to move. The cylinder is fixed to the frame of the machine, and is filled with a light oil whilst the piston is provided with a built-in adjustable valve that allows the oil below the piston to pass to the area above it slowly so permitting the saw frame to descend gently.

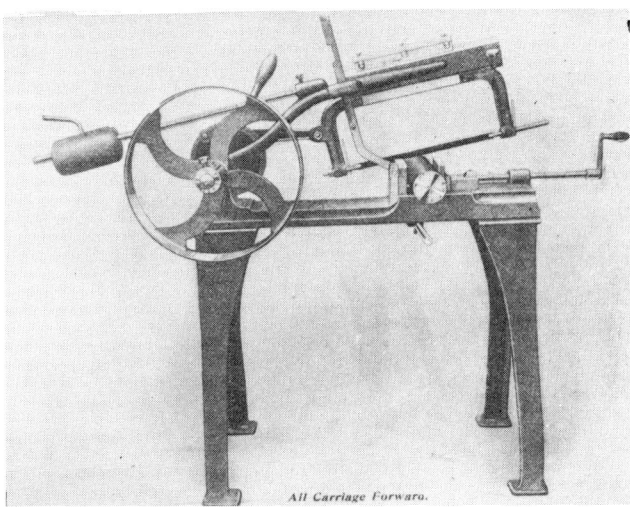

Fig. 13.6   Elementary power saw.

Fig. 13.7   The Cowell bench hacksaw.

Additionally the piston houses a comparatively large non-return valve, this is a valve that operates one way only, so designed as to allow the saw frame to be lifted quickly once the saw itself has completed its cut.

By combining the dashpot system with a pump, whose working stroke is timed to become effective when the saw frame is about to

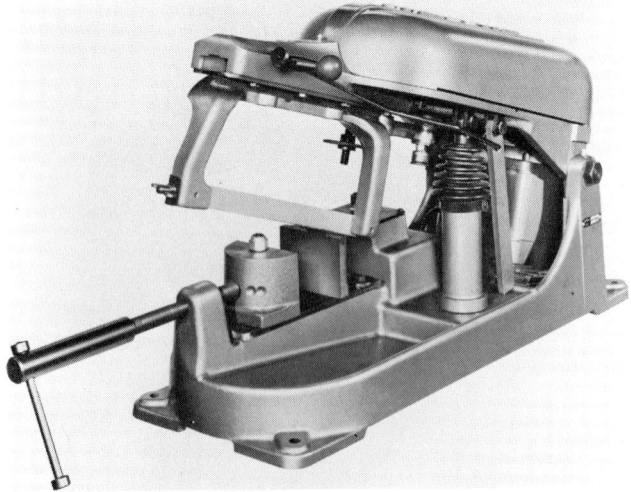

Fig. 13.8   The Pacera bench hacksaw.

make its return stroke, it is possible to provide a practical and efficient relief for the saw blade itself. Additionally, if the adjustable bleed valve in the dashpot can be arranged to close when required, the saw frame can be raised mechanically. In all the bench hacksawing machines illustrated, a switch is provided to cut off the electrical supply to the driving motor as soon as the saw has finished its cut.

Band Saws

Band saws used for metal cutting follow the lines employed with woodworking machines. The linear speed of the saw itself has to be adjusted to a rate suited to the sawing of metals. The saw tooth used is similar to that of the common hacksaw, the teeth being formed on a steel strip varying from $\frac{1}{4}$in. to $\frac{1}{2}$in. for the most part.

In the past, experiments with unconventional saw blades have been carried out. In one case the saw used had no teeth whatever, relying on the heat generated at the point of contact with the work to melt the metal immediately involved. In this way a clean cut was made whilst the cutting agent was but little affected, the heat generated being hardly communicated to it all. Though seeming to be quite a successful method it does not appear to have caught on generally.

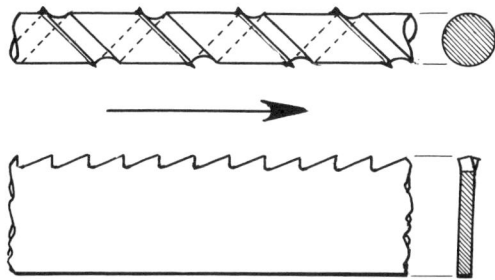

Fig. 13.9   Spiral blade saw compared with the normal metal cutting band saw.

Another blade form once tried was a continuous length of hardened wire having burrs formed on it to act as teeth. The author has had personal experience of this form of saw. Its small diameter was a great advantage when sawing curves of small radius, but, alas, the burrs quickly became blunted so the life of the saws was short.

The particular configuration of this saw in comparison with normal practice is depicted in the illustration Fig. 13.9.

H

Cold Sawing Machines

When large billets of metal need to be cut to length, or rough shaped before machining, the normal hacksawing machine will seldom serve. It is then that cold sawing equipment is brought into use. Basically the machine is a circular saw rotating at a suitable speed with means of gripping the work and feeding it towards the saw. Early machines followed the same configuration as the normal hacksawing machine. Instead of carrying a reciprocating saw, the frame carries a bearing assembly for the saw mounting as well as the worm gearing used to drive the saw.

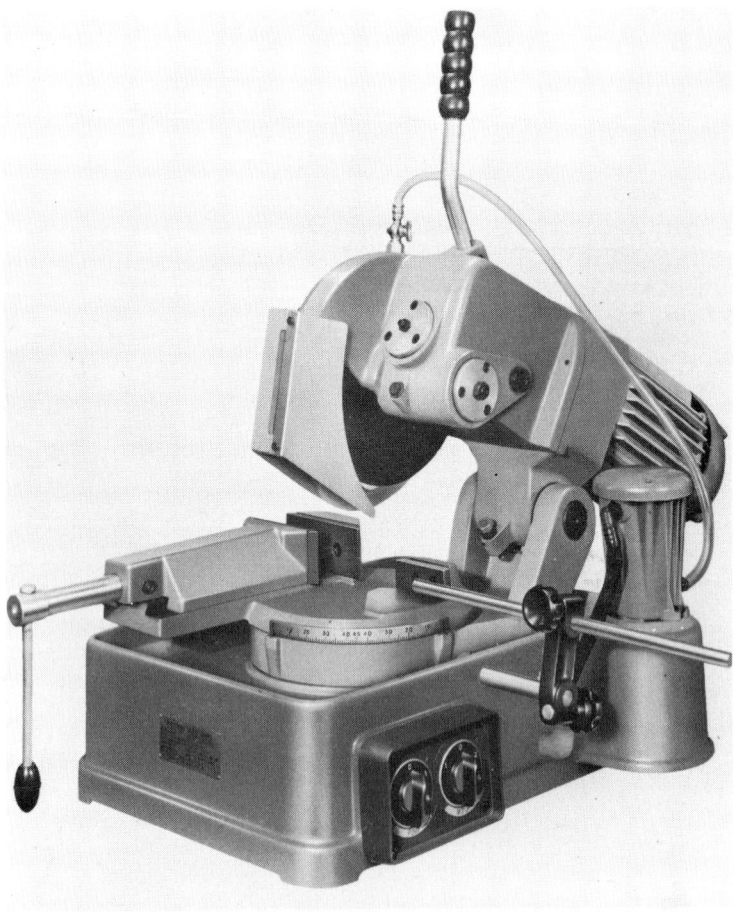

Fig. 13.10   The Pacera cold saw.

The equipment illustrated in Fig. 13.10 is a modern cold-sawing machine made by Messrs. W. J. Meddings of Slough. Designed for cut-off duty the vice may be angled in one plane only. Length stops are provided and also an independently driven electric suds pump, for cold saw working needs a plentiful supply of coolant for success.

Lathe Attachments

The advisability of providing either hacksawing or cold-sawing attachments for use on the centre lathe may be questioned by the professional machinist. Nevertheless, successful equipment has been made, generally by amateurs whose requirements must always envisage the use of the lathe, or indeed any other machine tool for purposes other than their designers might have contemplated originally.

A hacksawing attachment is illustrated in Fig. 13.11. This is virtually the same machine as that illustrated earlier as a complete

Fig. 13.11   Lathe hacksawing attachment.

Fig. 13.12   Lathe cold sawing attachment.

Fig. 13.13   Parts of the table assembly.

motor-driven unit. In this instance, however, the reduction gearing has been omitted, together with all bearing assemblies, the crankshaft being caught in the self-centring chuck. As may be seen, the machine bolts to the bed of the lathe, being held in place by a single large bolt passing between the bed shears through a clamp plate engaging tenons on the underside of the ways. No relieving device for the saw blade is provided, nor is the machine fitted with an automatic switch to cut off the power when the saw has completed its cut, as to provide this facility would have meant disturbing the electrical connections to the lathe itself.

Circular or cold-sawing attachments for the lathe have usually taken the form illustrated in Fig. 13.12. This consists of a table provided with angular and straight fences similar to those used in woodworking machinery. The table is mounted on supports bolted to the lathe cross slide which is locked when the table assembly has been set correctly in position. The parts of this assembly are shown in greater detail in the illustration, Fig. 13.13.

The saw itself is mounted on an arbor set between centres in the lathe, and driven at speeds suited to the work in hand. One of the advantages of the lathe attachments is that variation in spindle speed is available because it is already built into the lathe itself.

CHAPTER FOURTEEN

GRINDING MACHINES

GRINDING AS A METHOD of producing a finish to work as well as for sharpening hand tools appears to have been introduced about 1500. If we accept that the polishing of armour is a grinding operation then there is plenty of evidence that rotating abrasive stones were used for the purpose.

Leonardo da Vinci produced designs for several grinding machines, though it is not certain that they were ever made. However, he appears to have envisaged all the forms of grinding that are now normal practice.

From time immemorial natural stones and rocks have been used for the sharpening of tools and weapons of war. These abrasives vary widely in their performance a matter that is discussed fully in a book 'Sharpening Small Tools' published by Model & Allied Press Limited.

While the selection of suitable abrasives for hand use is a simple enough problem, the choice for machine grinding, at least in early times, was limited.

In the engineer's workshop the initial purpose of this operation was simply that of tool sharpening, in particular that of restoring a keen edge to lathe tools. The material originally used and made into grinding wheels was natural emery. A patent for the making of wheels from this substance being taken out as long ago as 1842. But little use seems to have been made of it until somewhere in the 1880's, by which time grinding was becoming an accepted method of machining components as well as sharpening tools.

For many years sandstone had been used, but this material, much favoured by woodworkers for grinding chisels and plane irons, needed to be flooded with water during the operation both to reduce wheel wear and to keep the tool cool. A typical sandstone wheel is illustrated in Fig. 14.1.

The machine consists of a framework supporting a pair of bearings that carry the wheel axle, and a water-tight trough surrounding one half of the wheel. Sometimes the trough was partially filled with water, at others a drip can was placed over the wheel to dispense a limited supply of water for cooling purposes.

The wheel itself was either hand-cranked, treadle-operated or driven from the workshop lineshafting as may be seen in the frontispiece illustration.

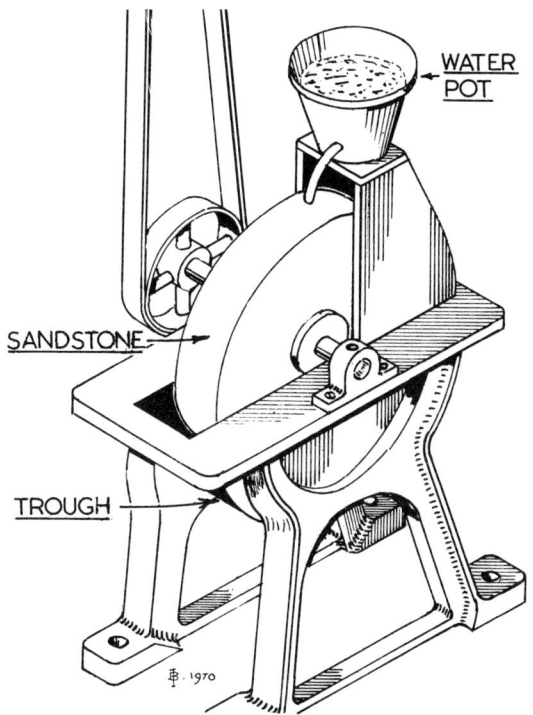

Fig. 14.1   The Sandstone.

The softness of the stone used lead to 'ringing' and deformation of the surface, and there were many attempts to combat this. One example is the illustration in Fig. 14.2. This is the double grindstone patented by Wm. Muir in 1853.

The stones were mounted on axles parallel with one another, and were made to rotate in contact with each other but at different speeds so that the scrubbing action resulting, kept the wheel surfaces flat.

Sandstone wheels are slow in operation, so it is not surprising that industry began to look around for means of increasing the rate at which tools could be ground. The use of natural emery has already been mentioned, and there is some evidence that wheels made from the material were in work about 1866, however, it was not until 1888 that emery wheels were in general use when a suitable bond for the abrasive grains had been found.

One disadvantage of natural stones and abrasive materials is that they are not of uniform quality, so operational difficulties were experienced when using them.

Fig. 14.2   The double grindstone.

The discovery in the laboratory of a method of producing pure corundum, an aluminium oxide similar to emery, provided the solution to the problem of non-uniformity. Together with silicon carbide, another abrasive material capable of artificial production, aluminium oxide is the basic material now used for abrasive wheels. There are now several makers of artificial abrasives each giving their product its own trade name.

Grinding Machines

In the past the tool grinding machine took the simple form depicted in Fig. 14.3 illustrating a small bench grinder intended principally for the amateur worker. The spindle ran in cast-iron bearings and simple tool rests were provided. The wheel mountings followed what has now become the common practice as illustrated in Fig. 14.4.

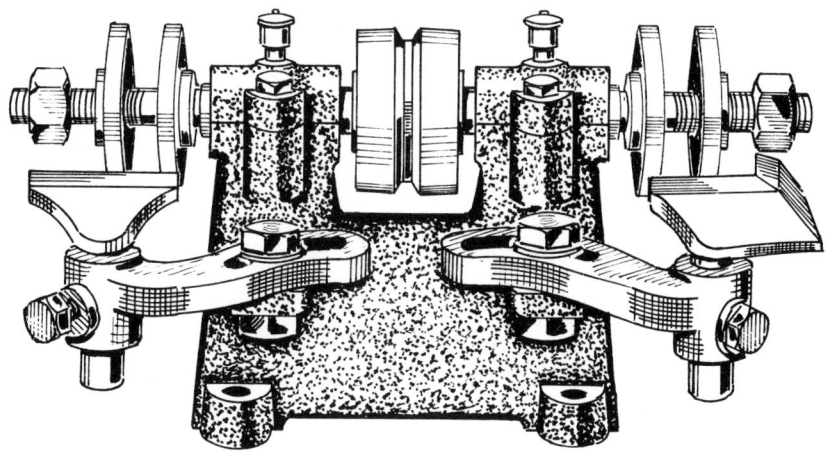

Fig. 14.3   The Simple bench grinder.

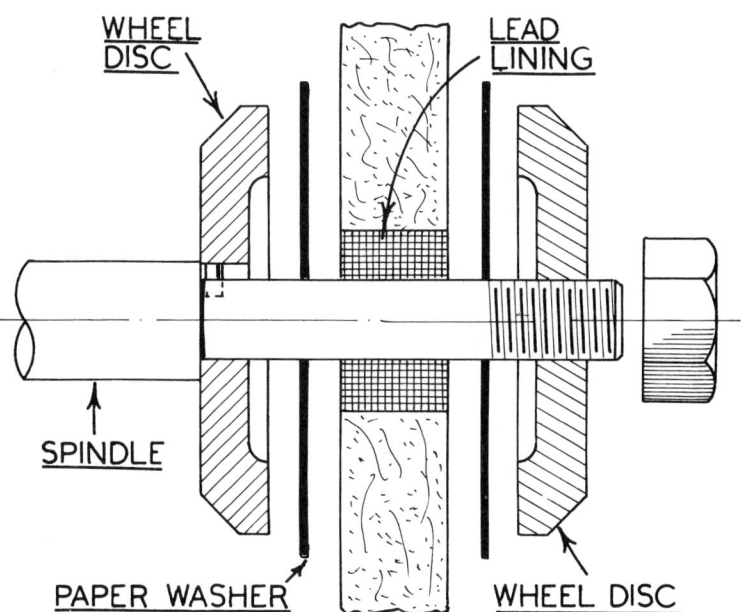

Fig. 14.4   Mounting the grinding wheel.

The arrangment consists of a pair of recessed metal discs, one of which is sometimes keyed to the machine spindle, and paper washers that are interposed between the discs and the sides of the wheel. These paper washers are an important item in the set-up as they permit the discs to bed against the wheel and grip it firmly.

Both emery wheels and those made from artificial abrasives need to be run at a high surface speed. In order to achieve this, a six-inch diameter wheel, for example, must be mounted on a spindle able to revolve at 3,000 revolutions per minute.

Naturally, and if continuously employed, the cast-iron bearing would not last over long. So, after an interim period when bronze bearings with adequate lubrication arrangements were employed, ball bearings have become the 'correct wear' for the modern tool grinder.

The advent of the electric motor in the machine shop eliminated the forest of belting by means of which the tools were once driven, and gave place to machines each with its own electric motor. The tool grinder was perhaps the simplest to modify, for all that was needed was a motor with an extended spindle and bearings capable of supporting the abrasive wheel correctly.

For the most part, electric grinders are made double-ended so that a pair of abrasive wheels can be mounted, one for roughing out the tool shape whilst the second is used for finish grinding. See Fig. 14.5.

Fig. 14.5   Electric bench grinder.

Angular Grinding Rests

Lathe Tool grinding machines in the past, have usually been provided with the simplest of rests; thus the operator has had to cant the tool in order to impart the clearances necessary beneath its cutting edges. Of late, however, there has been a development designed to provide a rest that can be tilted at definite angles in relation to the side of the grinding

Fig. 14.6   The Angular grinding rest.

wheel, in this way the correct clearance angles can be imparted without guess work.

A typical example is illustrated in Fig. 14.6 where a small bench grinder is seen with the angular rest fitted. The elementary angular rest shown needs the use of a protractor to measure or set the angle subtended by the rest with the side of the wheel. The rest depicted in Fig. 14.7 however, has a simple scale of degrees marked on the

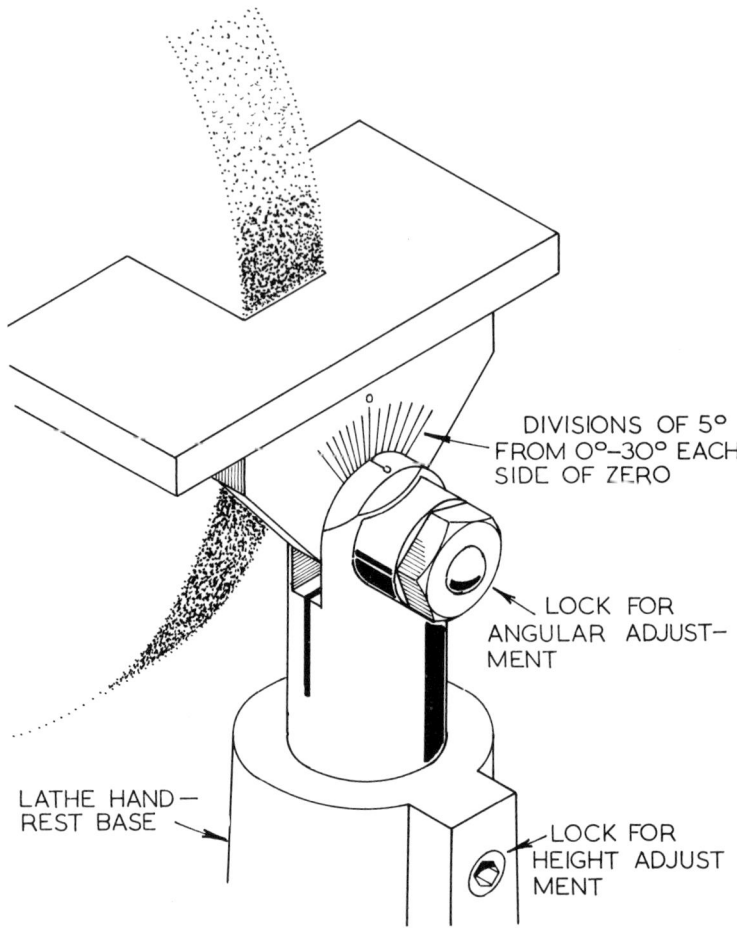

DIVISIONS OF 5°
FROM 0°–30° EACH
SIDE OF ZERO

LOCK FOR
ANGULAR ADJUST–
MENT

LATHE HAND –
REST BASE

LOCK FOR
HEIGHT ADJUST
MENT

Fig. 14.7   Angular grinding rest with built-in scale of degrees.

platform, each mark on it representing 5 degrees, to enable the user to
adjust the rest quickly.

The small grinder illustrated in Fig. 14.9 was made for the particular purpose of sharpening lathe tools made from tungsten carbide. In
order to do so correctly the direction of the abrasive grains in the wheel
must be downward from the cutting edge. The wheel itself must,
therefore, be capable of reversal and the tool ground on a rest at an
angle to the face of a cup wheel. If the direction of grinding is *not*

Fig. 14.8   Angular grinding rest being used.

downward then the cutting edge may be roughened and its durability and sharpness impaired.

Rough grinding of tools made from tungsten carbide is carried out with a special 'green grit' wheel. It is usually worked dry but the application of a little grease or oil to the wheel, though slowing up the

Fig. 14.9   Grinder for Tungsten Carbide tools.

rate of cutting, will produce a sufficiently fine finish for all practical purposes. In the professional workshop, however, diamond impregnated plastic or steel wheels are employed to impart a very high finish to the cutting edge.

A typical grinding machine for sharpening tungsten carbide tools in the professional workship is illustrated in Fig. 14.10. This machine, made by Messrs. Meddings of Slough, England, has tilting work tables and angular adjustable fences so that correct rake clearance and approach angles can be given to any tools being serviced.

Fig. 14.10 Grinder for Tungsten Carbide tools.

The machine, as will be seen, is double ended permitting tools to be rough-ground at one end then transferred to the diamond wheel at the other for final treatment.

Cutter Grinding Machines

The development of milling cutters of all types has lead to the production of grinding machines of some versatility. When one considers the variety of tools and cutters that need to be sharpened, and these include reamers, end mills and saws, then the grinders used need to be very versatile.

As with other machine tools the basic arrangements of early cutter grinders were founded on the elementary centre lathe. But advancement in cutter technology led to developments in the machines for sharpening them. These improvements virtually eliminated the characteristics of the centre lathe, resulting in the production of specialised machines capable of dealing with all types of cutter.

The fundamental difficulty with all machines used for grinding is the exclusion of abrasive dust from their moving parts. It was necessary, therefore, to take exceptional measures to prevent the dust and the metallic particles produced by the grinding operation itself. For the most part these measures comprise the covering of all machine slides, and labyrinth arrangements to protect spindle bearings and other shafts likely to be affected.

Grinding as a Machining Operation

We have examined the use of grinding for tool sharpening and must now consider the process as a method of component production. There are two advantages to be gained from grinding. In the first place the finish imparted to the work is superior to that resulting from other practices, except perhaps that of lapping which produces a superb surface finish but is very time consuming; secondly, it is of no importance whether the component is made from soft material or has been case-hardened to increase its wearing properties. Nor does the class of material inhibit the process provided the correct grade of abrasive wheel is used.

There are two basic grinding methods; these are cylindrical grinding and surface grinding. Cylindrical grinding is carried out on components either externally or internally. Examples of the former are shafts and mandrels for the mounting of work, whilst the latter is exemplified by the process commonly applied to the cylinders of automobile engines. though nowadays honing is gaining popularity because it is

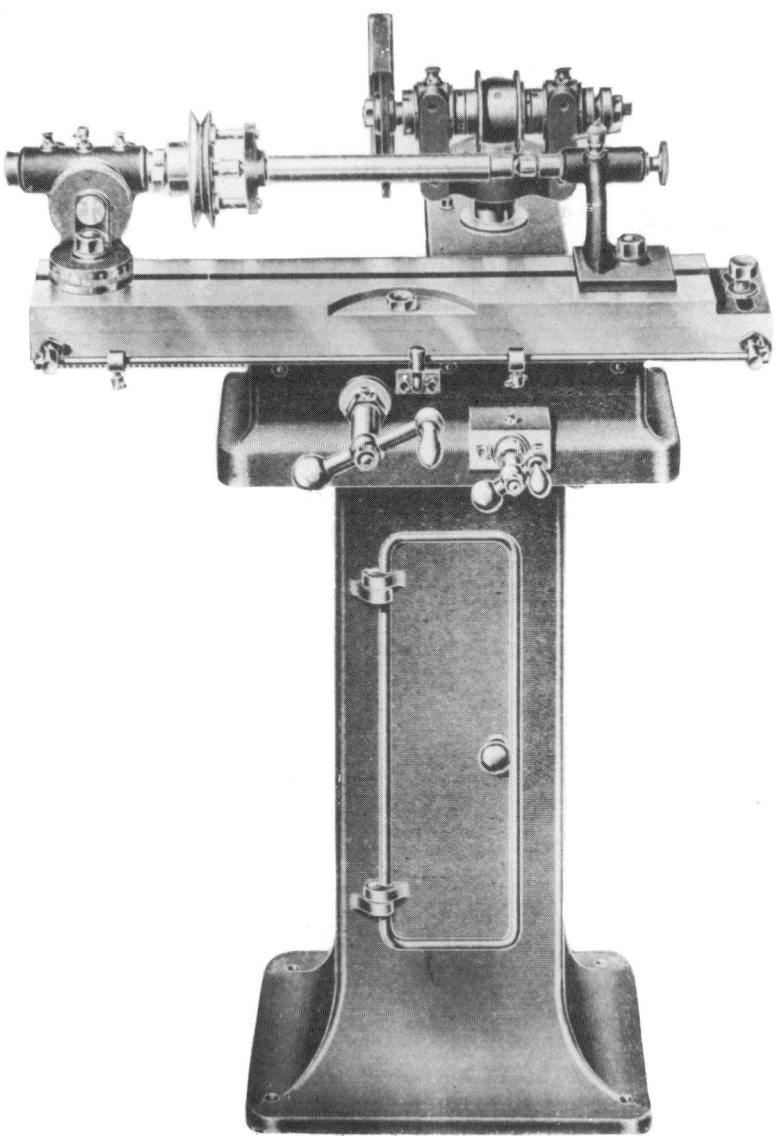

Fig. 14.11   Set-up for cylindrical grinding

somewhat easier to carry out and probably produces a more desirable finish to the cylinder wall.

Surface grinding is employed for the finishing of plane surfaces such as the beds of lathes, and small machine components suitable for mass grinding. The work is usually mounted on magnetic chucks that enable rapid loading and unloading of the parts themselves.

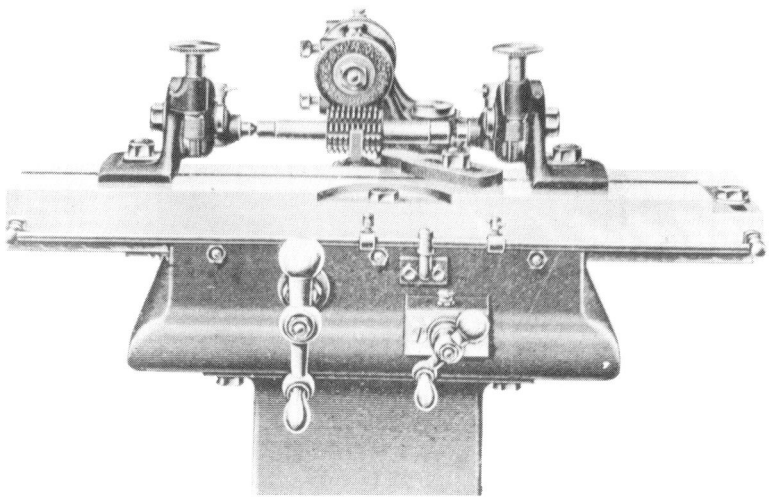

Fig. 14.12   A cutter grinder.

External grinding is typified by the illustration Fig. 14.13 at A. The work here is supported between centres and is passed across the face of the abrasive wheel whilst being rotated in the opposite direction to the wheel itself.

Internal grinding is performed in two ways. The first of these is illustrated in Fig. 14.13 at B. In this case the work is held stationary whilst the rotating grinding wheel itself describes an orbit that may be adjusted to suit the diameter of the work and the depth of cut required. By contradistinction in the second method, which is illustrated at C, the axis of the rotating abrasive wheel is held stationary whilst the work itself rotates. The wheel is passed backwards and forwards through the workpiece rotating, as before, in the opposite direction to the wheel.

The basic principle of this method is employed in the simple toolpost grinding equipment seen in the illustration Fig. 14.14.

A typical small cylindrical grinding machine is that depicted in Fig. 14.15. The equipment illustrated here is capable of both internal as well as external grinding operations. The internal grinding unit is mounted in a cradle set above the bearing housing of the main wheel head and may be swung down and locked in the working position seen in the illustration Fig. 14.16.

The workhead, seen to the left of the first illustration, is separately powered, the motor being capable of stepless variable speeds providing a work speed range for example from 100 to 900 r.p.m.

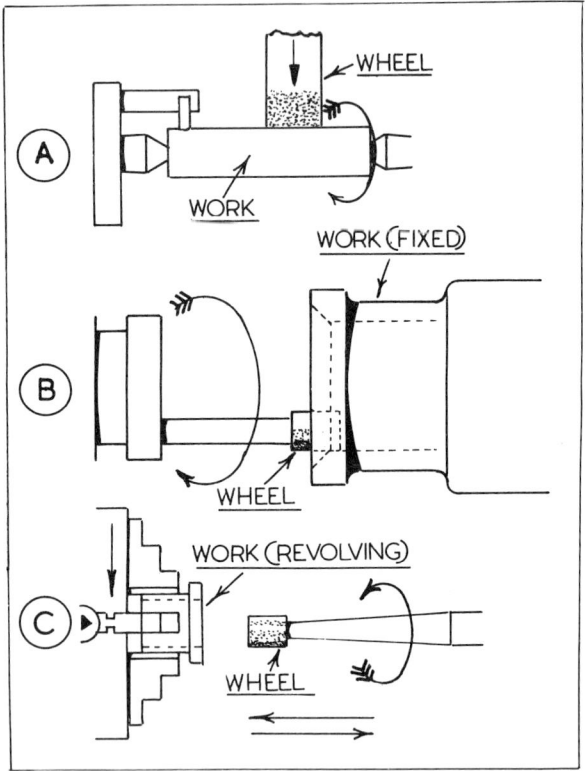

Fig. 14.13   Three methods of cylindrical grinding.

The workhead spindle is hollow and is provided with a draw rod so that equipment for dead centre or live spindle grinding can be mounted. The illustration, Fig. 14.17, depicts three methods of holding work. At 'A' a centre is mounted in the workhead spindle, at 'B' a collet is seen in the adapter, whilst at 'C' a self-centring chuck is attached to the spindle so that a wide range of work sizes can be accommodated for both external and internal grinding.

Machines of this type have either manually operated work table traverse, or are fitted with equipment that traverses the table hydraulicly, in which case the traverse rate is steplessly variable from zero to a maximum of 120 inches per minute.

Grinding machines need a copious supply of coolant; this is provided by a self-contained system consisting of a tank with filters and a circulating pump to feed the coolant to the wheel face. With so much suds available, and the product from the grinding wheel finely divided, the slides for the work table have to be fully protected to prevent ingress of water or abrasive material, particularly when the wheel itself is being dressed as must be done regularly.

Fig. 14.14   Toolpost grinding equipment.

Wheel dressing is usually carried out by means of an industrial diamond set in a fixture mounted on the work table, and able to be fed to the wheel by the advance and withdrawal mechanism of the wheel-head itself. When the required feed has been established the diamond is passed across the face of the wheel by the work table traverse.

In this connection it must be possible to adjust the work table auto traverse so that it makes a traverse and reverse cycle that will accomodate work of any length within the capacity of the machine itself. This is accomplished by the stops set in the T-slot seen above the large hand wheel on the front apron of the machine. These stops engage a

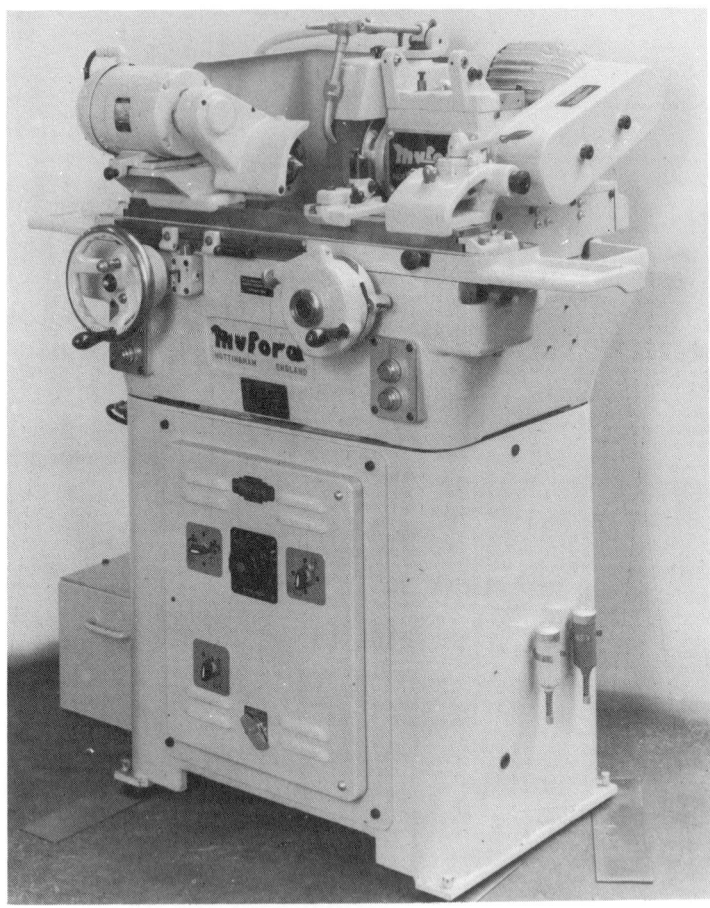

Fig. 14.15   The Myford Cylindrical grinder.

Fig. 14.16   The grinder with internal grinding unit in place.

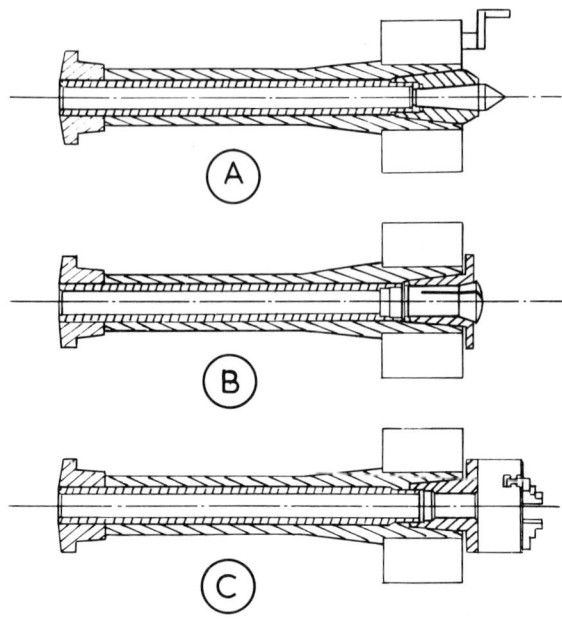

Fig. 14.17   Three methods of holding work in the grinder.

reversing device that operates either mechanically or hydraulicly according to the drive fitted to the work table.

Surface Grinding ˇ

There are two ways of surfacing grinding parts. In the first plaee they may be ground by the edge of the wheel whilst mounted on a magnetic chuck that passes back and forth under the wheel. In the second place the work is mounted on a rotating table provided with a magnetic chuck and is ground by a cup wheel rotating in contact with the work. This method is particularly suitable for the mass production of components that lend themselves to setting up on a magnetic chuck. Both methods of surface grinding are illustrated diagrammatically in Fig. 14.18.

Before leaving consideration of grinding practice, two particular techniques need to be mentioned. Both come under the heading of cylindrical grinding. The practice of centreless grinding has now been in vogue for many years and has the merit that, as may be gathered from its name, work is free to pass across the face of the abrasive wheel under the guidance of a control wheel and does not need to be supported on centres as previously described. In this way it is possible to deal with parts or stock material of considerable length.

The diagram, Fig. 14.19, shows the way in which the process is carried out. The work is supported on the rest seen in the illustration

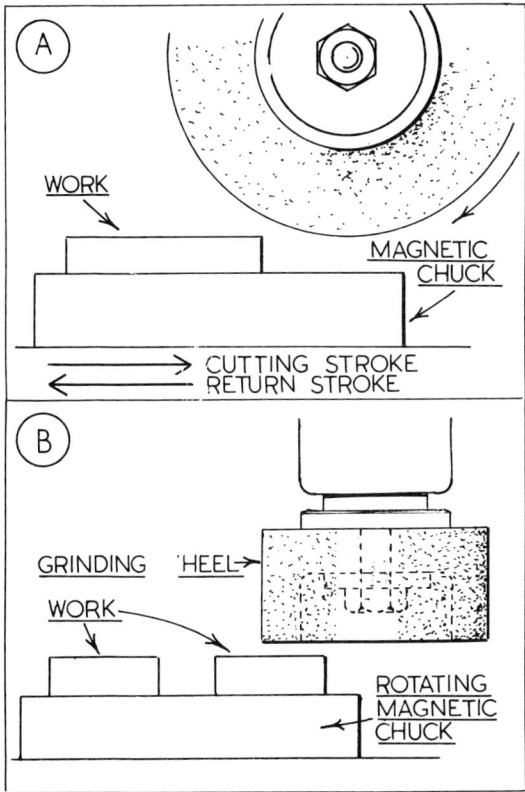

Fig. 14.18   Surface grinding methods.

and is rotated against the grinding wheel by the control wheel also depicted in the diagram. The depth of cut is determined by adjusting the distance between the centres of the grinding and control wheels.

The production of accurate screw threads has been much improved by the introduction of thread grinding. This is the process by means of which the threads themselves, and the taps and dies used to form them by hand methods, are generated by a grinding process.

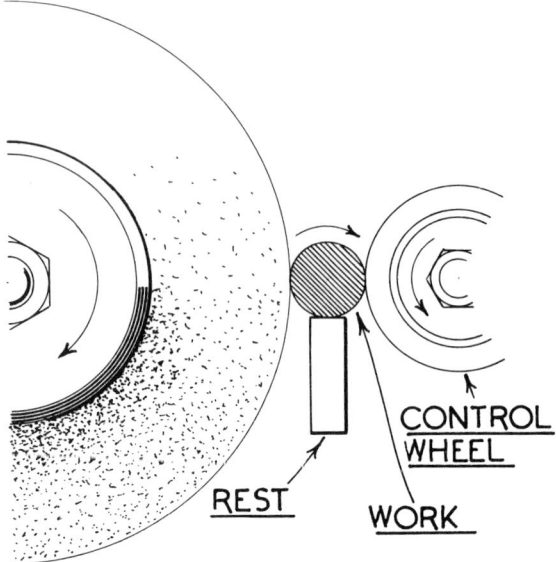

Fig. 14.19   Principles of centreless grinding.

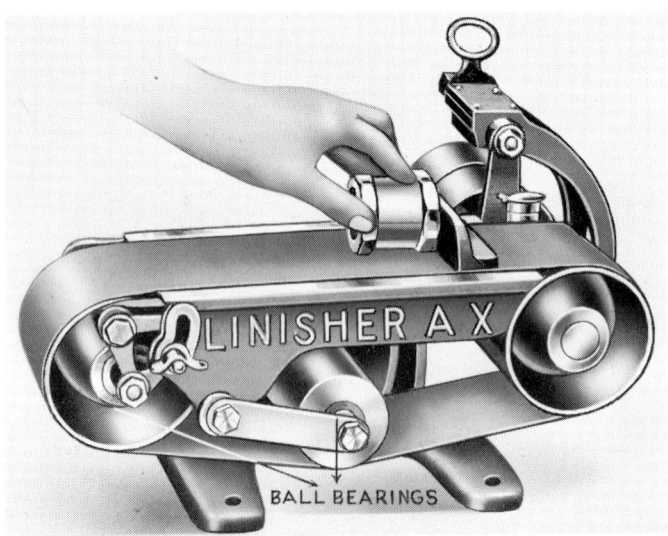

Fig. 14.20   The Linicher.

The Linisher

The piece of equipment illustrated in Fig. 14.20, cannot by any stretch of the imagination, be called a precision tool. As may be gathered from the illustration its purpose is, for the most part, to brighten the facets of nuts, the hexagon portions of brass pipe components, and any other work not requiring accurate dimensional finish.

As will be seen the linisher employs an endless abrasive band running over a pair of pulleys while a third and smaller pulley is used to take up any slack in the band resulting from operating the machine. This it does by moving freely of its own weight. Adjustment of the abrasive band is carried out by means of the left hand pulley which is capable of being moved and locked by the device seen in the illustration.

INDEX